Springer Proceedings in Physics 24

Springer Proceedings in Physics

Managing Editor: H. K. V. Lotsch

Recent Topics in Theoretical Physics

Proceedings of the first Nishinomiya-Yukawa Memorial Symposium, Nishinomiya, Japan November 8–9, 1986

Editor: H. Takayama

With 30 Figures

Springer-Verlag Berlin Heidelberg New York
London Paris Tokyo

Professor Hajime Takayama
Research Institute for Fundamental Physics, Kyoto University,
Kyoto 606, Japan

ISBN-13:978-3-642-73213-3 e-ISBN-13:978-3-642-73211-9
DOI: 10.1007/978-3-642-73211-9

Softcover reprint of the hardcover 1st edition 1988

2153/3150-543210

Foreword

My dear friends,

I am very pleased and honored to give the opening address in the first Nishinomiya-Yukawa Memorial Symposium on Theoretical Physics. Nishinomiya City wishes to extend a warm and sincere welcome to the many participants here in this Symposium.

Nishinomiya is the city where Dr. Hideki Yukawa (1907–1981) was living when he published the famous paper "On the Interactions of Elementary Particles. I" in 1935. For this work he was awarded the Nobel Prize for Physics in 1949.

To celebrate the 50th anniversary of his meson theory, our city has started the "Nishinomiya-Yukawa Memorial Activities" to promote the study of theoretical physics, which, we believe, is important for the progress of human society. The annual activities consist of this Symposium, the Nishinomiya-Yukawa Memorial Prize for promising young physicists and Memorial Lectures for citizens every year. They are conducted by the Steering Committee, the chairman of which is Professor K. Nishijima, the director of the Research Institute for Fundamental Physics, Kyoto University. I would like to express my sincere gratitude to the members of this Committee for their great efforts, as well as to the members of the Organizing Committee of this Symposium.

Dr. Yukawa said in his book "Tabibito" ("A traveler" in Japanese), "One who inquires into the truth is just like a traveler without a map." I shall be very happy and proud if this Symposium is successful and provides a useful map for many, particularly for young, physicists.

Thank you very much.

Yoneji Yagi
Mayor of Nishinomiya City

Preface

This volume contains lectures presented at the first Nishinomiya-Yukawa Memorial Symposium on "Recent Topics in Theoretical Physics" held on 8–9 November 1986 in the City Hall of Nishinomiya.

The purpose of the present Symposium was to bring together physicists active in different fields and to present introductory overviews of a high standard in several specialized topics that have undergone substantial development in the last few years.

During recent decades, rapid development and a branching into more and more specialized fields have hindered most physicists in understanding fields other than their own. We felt that this is largely due to a lack of understanding of basic concepts that are often buried in highly technical formalism. We have often experienced, however, that similar concepts appear repeatedly in various fields of physics, and that a knowledge of other fields somehow greatly stimulates the progress in our own field. Keeping this point in mind, the lecturers were asked to avoid technical details and to present key concepts of the developments, so that essential features would also be understandable to non-specialists.

Because of the limited time in the Symposium, we were forced to select only a few topics from each field of physics. We chose: superstrings, lattice gauge theory, the very early universe, gravitational collapse, problems concerning solar neutrinos, the quark cluster picture of nuclei, quantum Hall effects, Fermi surface effects, spin glasses, and the statistical physics of pattern formation.

The present proceedings, while they have an introductory character, consist at the same time of reports on the most recent developments in these areas. All the authors made an effort to meet these requirements, which we greatly appreciate. We hope that these proceedings will benefit physicists of many different horizons who are interested in current progress being made in fields other than their own.

The organizing committee of the Symposium consisted of:

M. Fukugita (Kyoto University)
M. Kobayashi (KEK)
A. Tomimatsu (Nagoya University)
M. Ichimura (University of Tokyo)
H. Takayama (Kyoto University)
A. Yoshimori (Osaka University)

The Symposium was organized under the auspices of the Education Board of Nishinomiya City, and the Research Institute for Fundamental Physics, Kyoto University.

Kyoto, June 1987 *Hajime Takayama*

Contents

Contents

Superstrings

K. Kikkawa

Department of Physics, Osaka University, Toyonaka 560, Japan

The string model, which appeared as a theory of hadrons in the early 1970s is now being considered as a promising candidate for a unification theory of all fundamental interactions. After surveying difficulties in attempts of unifying theories in local field theory, we describe why the superstring is considered hopeful and what issues remain unsolved.

1. Particle Hypothesis

Since Democritus (460-370 B.C.), we accepted for a long time the particle hypothesis, i.e., that the fundamental constituents of matter are indivisible and indestructible particles. In fact it has been quite successful, and indeed, no experimental contradiction nor any sign of scepticism to the hypothesis has been found. R.P. Feynman mentioned once that if he were allowed to communicate only one scientific fact to the next generation, he would tell them that our world was made of particles.

Nevertheless, we are not satisfied with the particle hypothesis. First of all we have too many fundamental particles. Particles which have been called elementary, such as the proton, the neutron, the pion and all other hadrons are no more considered to be fundamental. They are all constructed out of quarks. The number of quarks, however, is not small. According to the standard theory, we have at least three generations of fermions, each of which consists of at least 15 components, altogether 45 or more, as shown in Table 1. Bosons so far found, or believed to exist, are at least eight.

Table 1. At least three fermion generations as shown above exist. ν_R is not confirmed.

Fermions in the first generation — Bosons

Leptons — Quarks — Gauge bosons and Higgs

$$\left.\begin{matrix}\nu_L & \nu_R \\ e_L & e_R\end{matrix}\right\} 3\ [4] + \begin{bmatrix} u_L & u_R \\ d_L & d_R\end{bmatrix} \times 3 = 15\ [16] \qquad \left.\begin{matrix}\gamma\\ W^{\pm}\\ Z^0\end{matrix}\right\} 4 + \begin{bmatrix}\phi^+ & \bar{\phi}^0 \\ \phi^0 & \phi^-\end{bmatrix} = 8$$

Not all of these particles are necessarily independent. In the SO(10) grand unified theory (GUT) [1], for instance, each of the fermion generations consisting of 16 particles belongs to an irreducible representation of the group SO(10), and is considered a single particle having 16 possible different states, just as the nucleon has two different isospin states (the proton and the neutron). The idea of GUT unifies the particles and their interactions. The bosons in this model, however, cannot be so few, but 45 gauge bosons and 16 plus 45 Higgs particles have to be introduced to keep the symmetry.

As a price of unification we have to introduce a number of parameters in any GUT to break the symmetry because our world does not fully respect it. To get a satisfactory symmetry breaking, mostly by spontaneous symmetry breaking, one needs to introduce more than twenty adjustable parameters. The energy scale of the symmetry breaking is supposed to be of order of 10^{15} GeV to keep the proton stable, while observed fermions and bosons in Table 1 should be kept very light ($<$ 100 GeV). The fermion mass will be kept small if the theory has a chiral symmetry. Then how can the bosons be so light? Many people believe that the secret is the underlying supersymmetry, the symmetry between the boson and the fermion. This again requires a higher symmetry and more parameters.

We have pointed out that the energy scale of symmetry breaking is almost of order of the Planck energy ($\sim 10^{19}$ GeV). This suggests that the unifying theory may cover, not only the electromagnetic, the weak, and the strong interactions, but also the gravity interaction.

The theory of gravity due to Einstein is experimentally well established in large scale phenomena. The theory however cannot be quantized because it generates unrenormalizable divergences at every step of the calculation. There has been no self-consistent (well-defined) quantum gravity theory found and in fact one assumes that the graviton is described with a local field (the particle hypothesis).

Even if gravity is excluded, we do not believe that any unifying theory which suffers from so many complexities and parameters can be the "fundamental theory" of our world.

What are possible solutions to this problem? We consider three possibilities.

(a) The quarks and the leptons are not fundamental particles. There must be "more fundamental" particles --- the particle hypothesis.

(b) Our space-time is not as simple as we think. The dimensions and the structure of the space-time are more complicated [2] --- the geometry hypothesis.

(c) The fundamental constituent of matter is a divisible extended object such as a string, a membrane, etc. --- the extended object hypothesis.

As for (a) many models such as the technicolor and the preon etc. [3] have been proposed and analyzed in the past ten years. The situation has turned out to be hardly different from ordinary GUTs of quarks, and is in fact frequently more complicated. The geometry hypothesis (b) is not necessarily incompatible with the particle hypothesis. A number of attempts however have shown that the same troubles are inherited in so far as one sticks to the particle hypothesis. When the space-time structure is considered one cannot avoid the gravity theory. Attempts at the construction of a quantum gravity theory based on the geometry hypothesis have not been satisfactory notwithstanding much effort since the 1920s [2].

Now we depart from the traditional particle hypothesis, and consider the possibility of the extended object hypothesis (c). In the following, after making some observations on general extended objects, we report on the recent status of the superstring theory. The hypothesis of extended objects is intimately related to the geometry hypothesis. As will be described below, the quantum mechanics of strings works only in some special space-time dimensions or in special background geometry.

2. Why Strings?

Accepting the extended object does not necessarily mean the string. Historically the theory of string was being studied as a model of hadrons in the early 1970s. The string considered there was one with a spring constant of about 1 GeV/length. The quantum eigenstates generated by the rotation and vibration of the string were associated with hadrons. These eigenstates are now identified with quarks and gauge bosons provided that the spring constant is scaled up to the Planck energy ($\sim 10^{19}$ GeV). Any theory, if it is to be a unification theory, has to satisfy a number of conditions. Although the particle hypothesis is to be excluded above, we have to admit that the standard theory works as a good theory. The standard theory is a gauge theory in which zero mass quarks are interacting via a number of massless gauge bosons. The new theory therefore has to generate an effective gauge theory in low energy region (< several hundreds GeV) where massless gauge bosons and gravitons appear as particles. This is referred to as the massless condition.

Let us consider an n-dimensionally extended object, where n = 1 implies a string, n = 2 a membrane, etc. For simplicity let us assume that the object is uniform and is characterized by a single stress constant k_n, the dimension of which is $[k_n] = ML^{1-n} T^{-1}$ where M, L, and T represent mass, length and time, respectively. The theory has to be relativistic so that the light velocity c, $[c] = LT^{-1}$, will be involved. Suppose the object is rotating around its center of mass and generates the angular momentum (the spin) J, whose dimension is

$[J] = ML^2 T^{-1}$. Let us derive a relation between the spin J and the mass m (the total energy) of the rotating object by using dimensional analysis. The relation is unique in classical mechanics and is

$$J = A_n (k_n)^{-1/n} (mc)^{(n+1)/n} \tag{1}$$

where A_n is a dimensionless number [Fig. 1]. The relation shows that the zero angular momentum state with vanishing mass is possible for any n-dimensionally extended object.

In quantum mechanics the angular momentum is quantized and only states with integer (or half-integer) spins are allowed. The zero mass state in classical mechanics is no longer allowed because the zero-point energy of the object may shift the ground state energy. Namely, the mass m has to be replaced by

$$m^2 \longrightarrow m^2 + \Delta m^2 \tag{2}$$

in (1). The Δm^2 is well known as the Casimir effect. The spin-mass relation (the Regge trajectory) is now given by

$$J = A_n (k_n)^{-1/n} (mc)^{(n+1)/n} + \hbar B_n \tag{3}$$

where again A_n and B_n are dimensionless constants.

Consider the simple Nambu-Goto string defined by the action

$$S = -(k_1/4\pi) \int d^2\sigma \sqrt{-g}\, g^{ab} \partial_a X^\mu \partial_b X^\nu \eta_{\mu\nu} \tag{4}$$

where $X^\mu(\sigma^1, \sigma^1)$ represents the coordinate of the string and $g_{ab}(a, b = 0, 1)$ the metric tensor of world sheet swept out by the string, which is parametrized

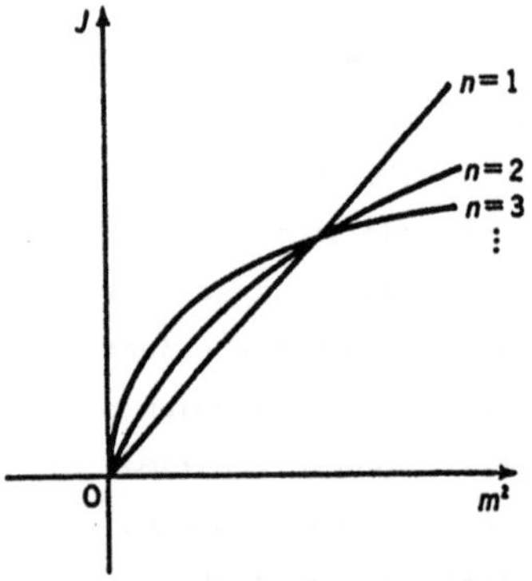

Fig. 1. Spin-mass relations in classical mechanics.

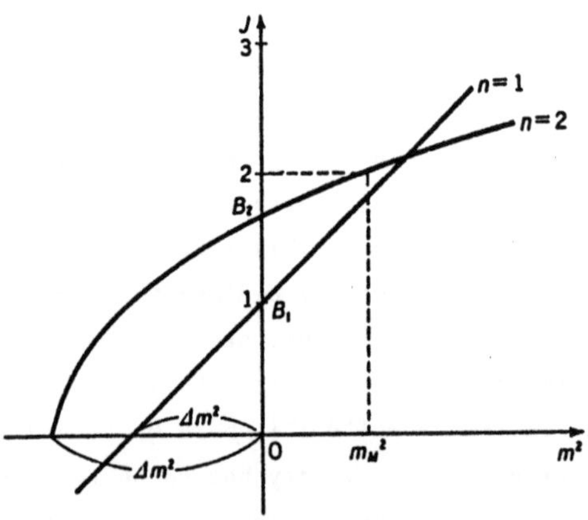

Fig. 2. Spin-mass relations in quantum mechanics.

by (σ^0, σ^1). [The string is parametrized by σ^1 at given time σ^0.] The space-time in which the string is embedded is assumed to have dimension D. The action S has significant symmetries; (1) the Lorentz invariance, (2) the reparametrization invariance, i.e., the invariance of S under $\sigma^a \longrightarrow \sigma'^a = f^a(\sigma)$ with $f^a(a=0, 1)$ being an arbitrary function, and (3) the Weyl symmetry, the invariance under $g_{ab} \longrightarrow \rho g_{ab}$ with ρ being an arbitrary function.

The spin-mass relation for this model [5] turns out to be

$$J = \frac{1}{p} \frac{c^2 m^2}{k_1} + \hbar \frac{D-2}{24} p \tag{5}$$

where the number p is either 1 or 2 depending on whether the string is open or closed, respectively, and D-2 represents the degrees of freedom of the string; the time and the longitudinal components of the string coordinates X^μ are not dynamical ones due to the reparametrization invariance of the action.

The massless condition discussed above requires that the intercept B_1 of (5) must be an integer (no fermion appears in this model), hence the D should be 2, 26, 50, $\cdots$. D = 2 is too small, and D > 26 are known to be excluded by the violation of unitarity. Although we have derived D = 26 from the massless condition, it is known that, unless D is 26, the Lorentz invariance is violated and also no consistent interactions between strings are possible. The dimensions in which a consistent quantum mechanics is realizable are referred to as the critical dimensions. In the superstring theory discussed later the critical dimension is 10.

What would happen for extended objects other than strings (n > 1)? Although no general quantum theory for n > 1 has been formulated, the author and Yamasaki [6] recently succeeded in calculating the intercept B_2 for a simple membrane model. The result can be stated as

$$B_2 = (\text{ a transcendental number }) \times (D - 3) \tag{6}$$

The intercept, therefore, cannot be an integer unless the space-time dimensions are transcendental, i.e. no massless state is expected in the membrane model. As for other cases (n > 2) we speculate that the situation would not be changed. The Weyl symmetry, which is possible only for n = 1, seems to play the crucial role in satisfying the massless condition.

In the string theory, massless particles with spin 1 and 2 (for closed string) always appear at critical dimensions. What is more, the particles with spin 1 and 2 were shown to behave as gauge particles and gravitons, respectively, by Scherk, Neveu, Schwarz and Yoneya [7]. This discovery motivated people to use the string for the unifying theory of all interactions including gravity.

3. Superstrings

Though the Nambu-Goto string is frequently used as a toy model for mathematical practices, it cannot be a candidate for the unifying theory because the spectrum does not contain fermions and the ground state turns out to be a tachyon with imaginary mass. The model construction having no defects has been initiated, again as a model of hadrons, by Neveu, Schwarz and Ramond (NSR) [8]. Mathematically self-consistent and tachyonless models having supersymmetry have been made from NSR by projecting out some subspace [9].

The superstring is a string made of spinning material. At each point on the string X^{μ} ($\mu = 0, \cdots\cdots, 9$) specified by a parameter σ^1 for given time σ^0, the two spinor fields $S^A(\sigma^0, \sigma^1)$ (A = 1 and 2) are associated, each of which is the Majorana-Weyl spinor in the critical dimension 10 and has eight independent components. The Weyl spinor means a massless field having a definite chirality (±1), while the Majorana spinor is a real field. These two properties are compatible only in 8n + 2 dimensions, hence in D = 10. The wave propagating along a string has two modes, a right-going (function of $\sigma^0 - \sigma^1$) and a left-going (function of $\sigma^0 + \sigma^1$), which define the orientability of string. Depending on the possible choice of combinations of the chirality of S^A and the left- and right-modes, one can construct a number of self-consistent models [10] as shown in Table 2.

SST-I is made of spinors with a definite chirality (either + 1 or -1 for both A = 1 and 2) and is left-right symmetric. The string is therefore non-orientable. Both closed and open strings are allowed.

In SST-II (A and B) both the left and right modes are independently incorporated, hence they are orientable. The type A is a non-chiral model having the chirality +1 for S^1 and -1 for S^2, or vice versa, which is not suitable for a realistic model. The type B is a chiral model i.e. S^1 and S^2 have the same chirality. Both A and B types consist of closed strings only. The heterotic string will be explained later.

Table 2 Superstring models in 10 dimensions

Models	Structure	Gauge particles	Supersymmetry
SST-I SO(32)	open+closed	496	Yes
-II A	closed	0	Yes
-II B	closed	0	Yes
Heterotic SO (32)	closed	496	Yes
-E(8)xE(8)	closed	496	Yes
-O(16)xO(16)	closed	240	No

The good properties of these models are all attributed to the high symmetry of superstrings. When a string sweeps out space-time, a two-dimensional manifold (world sheet) is drawn. The coordinates X^{μ} and S^{Aa} are considered to be distributed on the world sheet. The models have following symmetries.

(a) The Poincare invariance in the 10 dimensional Minkowski space.
(b) The reparametrization invariance of the world sheet.
(c) The Weyl invariance, an invariance under changing the scale of the two-dimensional world sheet metric.
(d) The supersymmetry under a transformation between X^{μ} and S^{A}.

The invariance (a) is obvious. (b) is necessary to eliminate negative normed states (ghosts) which arise from the time component of the coordinate. (c) plays an essential role as the conformal invariance when a gauge is fixed to the reparametrization, without which interactions between strings are unable to be incorporated. The critical dimension D is also determined by requiring that the conformal invariance should not be violated after quantizing the system (or anomaly-free condition of the conformal translation). (d) is necessary, not only for the boson-fermion symmetry but also for eliminating tachyons and for making the theory finite.

The superstring theory is frequently referred to as a miraculous theory. This is because of the anomaly-cancellation mechanism incorporated in the theory.

In field theories, when an action with certain symmetry is given, a quantized version of the theory does not always respect the symmetry. The well-known example is the chiral symmetry in a massless quark theory, in which the action is invariant under the chiral transformation of quark field ψ

$$\psi \longrightarrow \exp[i\alpha\gamma_5]\psi, \quad \bar{\psi} \longrightarrow \bar{\psi}\exp[i\alpha\gamma_5]$$

The Noether theorem superficially tells us that the axial current should be conserved. In reality, however, the current is not conserved. The reason is that in quantum theory the path integral measure in

$$\int d\psi d\bar{\psi} \exp[iS_{ch}]$$

is unable to keep the chiral invariance [11]. This can also be confirmed by carrying out the usual perturbation calculation. Those terms which violate the symmetry are called the quantum mechanical anomalies.

In some cases the anomaly destroys the whole theory. In gauge theory if the anomaly for the gauge symmetry existed, neither the unitarity nor the renormalizability would be guaranteed. Anomalies in field theory have been recently well studied in most cases where they arise. In the standard electroweak theory the fermion generation in Table 1 is formed so nicely that the anomaly for gauge invariance is cancelled altogether. In higher dimensions the principle of general coordinate invariance as well as the gauge invariance are generally violated by anomalies, unless some conditions are satisfied.

Let us return to the superstring. The superstring generates massless gauge bosons and gravitons according to the mechanism mentioned before, hence effectively a certain supersymmetric gauge theory and supergravity theory. The supergravity means the theory in which a spin 3/2 particle is incorporated as a super partner of the graviton. Here again anomaly should not be allowed to appear to keep the consistency of theory. In 1983 Alvarez-Gaumé and Witten [12] showed that SST-II model does not contain an anomaly, but that SST-I does. While SST-II has too small symmetry to describe our world, the SST-I can accommodate any symmetry, hence their discovery was taken quite seriously. In 1984, however, Green and Schwarz discovered that the anomaly is cancelled even in SST-I if and only if SO(32) gauge symmetry is incorporated [13]. They also pointed out that, if one could form a superstring with E(8) x E(8) symmetry, which is preferable from the phenomenological point of view, the model should show no anomaly. The model with E(8) x E(8) was constructed at once by the Princeton group [10]. This and the variations of it are called heterotic strings [Table 2].

The heterotic string has a quite artificial feature. The space where the string is embedded consists of the ten-dimensional Minkowski space M_{10} times a sixteen-dimensional torus T_{16}, the latter of which serves for generating the internal gauge symmetry E(8) x E(8) or SO(32). The string is closed and consists of two parts. One is the superstring part which moves in M_{10} only, and the wave always propagates rightwards along the string. The other is the bosonic (Nambu-Goto type) string which moves in both M_{10} and T_{16}, and the wave propagates leftwards along the string. All of these artificial structures are needed in order to make the model anomaly-free, conformal invariant, tachyonless, etc.

The heterotic O(16) x O(16) has a similar structure as above except for the space structure $M_{10} \times D_{16}$ instead of $M_{16} \times T_{16}$, where D_{16} is called an orbifold defined as $D_{16} = T_{16}/Z$ with Z being a certain discrete group. Readers are suggested to refer to ref. 10 for details.

The anomaly, although it was previously considered as an unsound feature, is now regarded as a substantial feature of quantum theory, and the anomaly-free condition is taken advantage of for finding new theories. In models listed in Table 2 all anomalies are cancelled by mathematical "miracles".

Another important property in superstrings is the finiteness of the theory. In field theories the divergences appearing in perturbation series become less as the symmetry of the action becomes higher, and eventually renormalized out. In the gravity theory it is not so even if supergravity symmetry is incorporated. In the superstring theory quantum gravity (the graviton) appears as a quantum mode of the string, and the gravity theory obeying the principle of general relativity is generated as an effective theory in a large scale limit where the spring constant is taken to be large. The divergence in string amplitudes is shown to be eliminated under the same condition of the anomaly cancellation [14]. It should be stressed that the superstring is the only available theory of quantum gravity.

4. Compactification

The critical dimension being 10 rather than 4 is the biggest problem in the superstring theory. In theory which includes gravity, however, the dimensions of space-time need not be four from the outset, since the structure of space is to be determined as a consequence of the equation of motion. The solution having the structure of the four-dimensional Minkowski space times a six-dimensional compactified space should be looked for in the model. So far no one has solved the equation of the string field. There are however a number of ways to find the classical structure of space-time, i.e., the vacuum structure.

One of the methods is to exploit the conformal symmetry of the theory. To see the idea let us take the Nambu-Goto model again. The action is given by (4) which governs the propagation of a string in Minkowski space. But when interactions of strings are taken into account, string fields may condensate to form a curved space metric just as the spontaneous symmetry breaking occurs in ordinary field theory. Then the space-time metric would be effectively replaced by $G_{\mu\nu}$ in (4) in place of $\eta_{\mu\nu}$. The new action is

$$S = -(k_1/4\pi) \int d\tau d\sigma \sqrt{-g}\; g^{ab}\partial_a X^\mu \partial_b X^\nu G_{\mu\nu}(X) + \cdots \tag{7}$$

where $\cdots$ means some term irrelevant to the metric. The action (7) is regarded as a non-linear σ-model on the two-dimensional manifold. If the parameter k_1 is large, the quantum amplitudes can be calculated in perturbation expansions in $1/k_1$. In general the anomaly for the conformal invariance appears if the metric $G_{\mu\nu}$ is not specified. In fact if the anomaly term obtained is required to vanish, one gets a set of Einstein-like equations for $G_{\mu\nu}$, which should determine the vacuum metric. The method is legitimate because the conformal invariance should be kept in any circumstance [20].

Although their method is slightly different from the one we have explained, Candelas et al. [15] found an attractive solution in the heterotic string with E(8) x E(8), which has the following characters.

(a) Space-time is compactified into the four-dimensional Minkowski space times a six-dimensional compact space K_6 called the generalized Calabi-Yau space.

(b) In the four-dimensional Minkowski space the solution exhibits the gauge group E(6) x E(8) with an approximate supersymmetry.

(c) Four generations of massless fermions appear.

All of these properties are favorable but not enough. In string theory there are only two parameters involved, the spring constant and the coupling constant in the string interaction. The former is fixed from our knowledge of the gravitational constant. The latter is related with the dilaton vacuum expectation value and is determined dynamically from the stability condition of the vacuum. No new parameters are allowed to be introduced, otherwise the conformal invariance is violated. The solution we want should provide all the correct low-energy data, including the cosmological constant, the neutrino mass and so on, without introducing any parameter. This is obviously one of the biggest issues in the superstring theory.

Recently many people are trying to find new models which have four, rather than ten, dimensional Minkowski space plus extra compact space [16]. This is again an issue requiring futher study.

5. Conclusion

There are many issues remaining in string theory. Possible new models for the unification theory should be sought. Even if a model, say E(8) x E(8) heterotic string, is promising, a compactified solution which describes our world has to be found. The research has just begun.

At this stage, the superstring model for the unification theory is supported by theoretical reasoning rather than experimental results. This is the only theory which provides a quantum theory of gravity. This is the only theory which generates no anomalies nor divergence. These have been all long-standing stumbling blocks in quantum local field theory.

Apart form the application to unification theory, the study of the physical and mathematical properties of strings is no less important. The string theory has to be reformulated as a field theory in space-time, i.e., as a second quantized theory. So far,three formulations have been proposed. The first one, proposed by Kaku and the author [17], is a formalism in the light-cone gauge. Although the theory is simple and useful in the perturbative spproach, a non-perturbative calculation cannot be done easily, since the relativistic covariance is not manifest there. The second formalism by Witten [18], which is manifestly covariant, has not been completed for the closed string. The third one proposed by Kyoto group [19] is manifestly covariant but it contains a set of extra parameters whose roles are not yet understood.

The theory of strings will be theoretically completed only when a field theoretical formalism is accomplished, otherwise non-perturbative studies will not be possible. The issues of compactification, and many others are non-perturbative problems.

We have encountered a lot of difficult problems in the string theory, but we have found a number of miraculous properties which are not shared by the local field theory. The theory of strings has greatly expanded the applicability of quantum theory. It was in the 1970s that the local field theory, proposed in 1932 by Heisenberg and Pauli, flourished. We should continue to work patiently in search of a conclusion.

References

1. See for example, G.G. Ross, Grand Unified Theories, Benjamin-Cummings (1985).
2. For instance, T. Kaluza, Sitz. Ber. Preuss. Akad. Wiss. K1 (1921) 966; O. Klein, Z. Phys. 37 (1926) 895.
3. E. Farhi & L. Susskind 74 (1981) 277; H. Harari, Phys. Lett. C 104 (1984) 159.
4. Dual Theory, ed. M. Jacob, Phys. Reports reprint Vol. I (North-Holland 1974).
5. L. Brink and H.B. Nielsen, Phys. Lett. B45 (1973) 332.
6. K. Kikkawa & M. Yamasaki, Prog. Theor. Phys. 76 (1986) 1379.
7. A. Neveu & J. Scherk, Nucl. Phys. B36 (1972) 155; J. Scherk & J.H. Schwarz, Nucl. Phys. B81 (1174) 118; T. Yoneya, Prog. Theor. Phys. 51 (1974) 1907.
8. A. Neveu & J.H. Schwarz, Nucl. Phys. B31 (1971) 86; P. Ramond, Phys. Rev. D3 (1971) 86.
9. J.H. Schwarz, Phys. Rep. 89 (1982) 233; M.B. Green, Surveys in High Energy Physics 3 (1983) 127.
10. D. Gross, J.A. Harvey, E. Martinec & R. Rohm, Nucl. Phys. B256 (1985) 253; L. Dixon & J. Harvey, Nucl. Phys. B274 (1986) 93; L. Alvarez-Gaumé, P. Ginsparg, G. Moore, and C. Vafa, Phys. Lett. 171B (1986) 155.
11. K. Fujikawa, Phys. Rev. D21 (1980) 2848.
12. L. Alvarez-Gaumé & E. Witten, Nucl. Phys. B243 (1983) 269.
13. M.B. Green & J.H. Schwarz, Phys. Lett. 149B (1984) 117.
14. M.B. Green & J.H. Schwarz, Phys. Lett. 151B (1985) 21.
15. P. Candelas, G.T. Horowitz, A. Strominger and E. Witten, Nucl. Phys. B258 (1985) 46; D. Nemeshansky and A. Sen, Phys. Lett. B178 (1986) 365.
16. K.S. Narain, Phys. Lett. 169B (1986) 41; H. Kawai, D.C. Lewellen, S.-H.H. Tye, Phys. Rev. Lett. 57 (1986) 1832.
17. M. Kaku & K. Kikkawa, Phys. Rev. D10 (1974) 1110 & 1823; E. Cremmer & J.L. Gervais, Nucl. Phys. B76 (1974) 209.
18. E. Witten, Nucl. Phys. B268 (1986) 253.
19. K. Itoh, T. Kugo, H. Kunitomo and K. Ogawa, Phys. Lett. 72B (1986) 186 & 195.
20. C. Callan, E. Martinec, M. Perry & D. Friedan, Nucl. Phys. B262 (1986) 593; E. Fradkin & A. Tseytlin, Phys. Lett. 158B (1985) 316; Phys. Lett. 160B (1985) 64; K. Kikkawa & M. Yamasaki, Prog. Theor. Phys. Suppl. No. 85 (1985) 228.

Quantum Chromodynamics on a Lattice

A. Ukawa
Institute of Physics, University of Tsukuba, Ibaraki 305, Japan

Current understanding of the dynamics of strong interactions based on quantum chromodynamics is reviewed with emphasis on the lattice approach. Recent results of numerical simulations on the hadron mass spectrum and the finite temperature behavior of QCD are discussed.

1. Introduction

The theory of strong interactions started in 1934 when Yukawa[1] proposed that the force between nucleons was mediated by a new particle having a mass of about 200 times that of the electron. This particle, now known as the pion, was discovered in 1947[2]. The number of strongly interacting particles, hadrons, has greatly increased since then, and several hundred of them are now known[3]. It is quite natural to suppose that so many particles can not be all elementary. In fact the observed pattern of hadrons can be very nicely explained if, following Gell-Mann and Zweig[4], one assumes that the hadrons are made up of a few kinds of spin 1/2 Dirac particles called quarks. According to the quark model the mesons having integer spins are made up of a quark and an antiquark, whereas the baryons having half-integer spins are composed of three quarks. At present five kinds(flavors) of quarks are needed to account for the known hadrons and at least one more is expected from theoretical considerations.

It has long been a puzzle that quarks, so useful for classifying hadrons, have never been directly observed in experiments. If quarks are real, this implies that the force binding quarks is sufficiently strong to prevent their escape from within hadrons. For some time the nature of the mechanism leading to such a 'confinement of quarks' was a mystery. For instance, the conventional mechanism of particle exchange is clearly inadequate since it typically leads to a force with exponential or power decrease at large separations which obviously does not confine. It is now widely believed that non-Abelian gauge field theories[5] possess a natural mechanism for confinement. These theories are a generalization of quantum electrodynamics (QED), the well-known theory that describes the interaction of electromagnetic fields and charged particles. Instead of the single commuting charge of QED, these theories have a number of charges Q^a, $a=1,\ldots,n$, which form a Lie algebra $[Q^a,Q^b]=if^{abc}Q^c$ of some compact non-Abelian (i.e., non-commutative) classical group where f^{abc} is the structure constant. This apparently natural generalization leads to a completely different dynamics; the 'electric' fluxes that emanate from the charges in non-Abelian theories are believed to be squeezed into a string-like configuration with a constant energy density per unit length. One thus has confinement of non-Abelian charges is such theories; any excitation that has a non-vanishing non-Abelian charge has an infinite energy and cannot exist in the physical spectrum.

Quantum chromodynamic (QCD)[6] is the non-Abelian gauge field theory of quarks based on the group SU(3). The SU(3) charge of the theory is called color. The quarks transform under the 3-dimensional fundamental representation and have three color charge states. The gauge fields come in eight colors and, being coupled to the color of quarks, are the agent of the force between quarks. Because of the confining property of the force carried by the gauge field, the quarks and more generally any excitation that has non-vanishing color are confined in QCD. The hadrons are natural exceptions since a quark and an antiquark combine to form a color singlet object (mesons) and so do three quarks (baryons). The property of non-Abelian gauge fields which is intimately connected with confinement is that the magnitude of its coupling strength becomes weaker at shorter length scales[7]. This property, called asymptotic freedom, is precisely the one needed to explain the so-called scaling phenomena of high energy hadronic interactions. Thus QCD accounts for both low (quark confinement) and high (scaling) energy dynamics of strong interactions and is now widely believed to be the theory of strong interactions.

The purpose of the present talk is to explain why quarks are naturally confined in QCD and how QCD allows quantitative calculations of the observables of strong interactions. The dynamics of confinement in QCD is not simple and the current understanding depends on several key ideas and methods. The first is the formulation of gauge theories on a lattice (with imaginary time)[8]. This not only provided a mathematically rigorous definition of QCD, but also established a close connection between the quantum theory of gauge fields and statistical mechanics. This latter aspect is particularly important. It allowed the application of statistical-mechanical methods which in turn showed the importance of the concept of phase transitions for understanding the dynamics of confinement. The second key was the application of numerical simulation techniques[9-11]. This has yielded a great deal of information on the dynamics of QCD which so far cannot be obtained by analytical methods. The numerical simulation is not simply a useful tool but at the moment the only method which allows extraction of quantitative predictions of lattice QCD.

We shall start this review with the lattice formulation of QCD (Sec. 2). We shall see how confinement naturally arises in this framework. The numerical simulation techniques are introduced in Sec. 3 and in Sec. 4 we describe how one calculates hadron masses in lattice QCD. We then examine the question of what might happen at high temperatures in Sec. 5. The last Sec. 6 provides a brief summary and an outlook on lattice QCD studies.

2. Lattice QCD and Quark Confinement

The lattice formulation of QCD requires a discretization of the space-time continuum. We shall always work with the imaginary time $x_4=it$. Thus the original space-time is the four-dimensional Euclidean space R^4. The choice of the lattice is in principle arbitrary. We shall use the simplest choice, namely the simple cubic lattice having the lattice sites at $x=na$ with $n=(n_1,n_2,n_3,n_4)$ an integer four vector and a the lattice spacing which sets the unit of length on the lattice.

The basic ingredients of QCD are the quark field and the SU(3) gauge field. The quark field $q_n^{i\alpha}$ is defined on each lattice site n and has the color index i=1-3 as well as the Dirac spinor index α=1-4. The gauge field $U_{n\mu}$ is defined on each link, namely the directed line connecting the site n and its nearest neighbor site $n+\hat{\mu}$ with $\hat{\mu}$ the unit vector in the μ-direction (μ=1-4). The SU(3) nature of the gauge group is encoded in the assignment that the field variable $U_{n\mu}$ is an SU(3) matrix which is a 3×3 complex matrix satisfying $UU^\dagger=U^\dagger U=1$ and $\det U=1$.

Having introduced the fields, we now define the action. For the gauge field, the simplest choice[8] is given by

$$S_g = \frac{1}{g^2} \sum_{n,\mu,\nu} \mathrm{tr}(U_{n\mu} U_{n+\hat{\mu},\nu} U^\dagger_{n+\hat{\nu},\mu} U^\dagger_{n\nu}) , \tag{1}$$

where g denotes the coupling constant. Geometrically, the expression inside the trace is the ordered product of four U's around an elementary square (plaquette) of the lattice. The relation between the lattice gauge field $U_{n\mu}$ and the more conventional field $A_\mu(x)$ is given by $U_{n\mu}=\exp(iaA_\mu(na))$. Using this, it is an easy exercise to show that (1) reduces to the well-known continuum form $\int dx^4 \mathrm{tr}F^2_{\mu\nu}/2g^2$ as the lattice spacing $a \to 0$ where $F_{\mu\nu}=\partial_\mu A_\nu - \partial_\nu A_\mu - i[A_\mu, A_\nu]$ represents the field strength. For quark action, the naive discretization of the Dirac operator gives

$$S_q = a^3 \sum_{n,\mu} (\bar{q}_n \gamma_\mu U_{n\mu} q_{n+\hat{\mu}} - \bar{q}_{n+\hat{\mu}} \gamma_\mu U^\dagger_{n\mu} q_n) - m_q a^4 \sum_n \bar{q}_n q_n . \tag{2}$$

with m_q the quark mass. Unfortunately, the action (2) has the undesirable feature called species doubling. This is intimately related to the periodic structure of the lattice and cannot be removed unless one explicitly breaks the chiral symmetry [12]. There exist several proposals[13] for the necessary modification of (2). The details are not needed here and are hence omitted.

The final step in defining QCD on a lattice is the quantization. This is done by the Feynman path integral: the vacuum to vacuum transition amplitude is given by

$$Z = \int \prod_{n\mu} dU_{n\mu} \prod_n dq_n d\bar{q}_n \, e^{-S_g - S_q} , \tag{3}$$

and the matrix element of an observable $\mathcal{O}$ by

$$\langle \mathcal{O} \rangle = \frac{1}{Z} \int \prod_{n\mu} dU_{n\mu} \prod_{n} dq_n d\bar{q}_n \, \mathcal{O} \, e^{-S_g - S_q} . \quad (4)$$

In these equations, dU denotes the invariant measure over the group SU(3). The quark fields q and $\bar{q}$ are anti-commuting Grassmann numbers and $dqd\bar{q}$ the corresponding measure.

QCD thus defined has several important properties. First of all it possesses the invariance under the local SU(3) transformation given by

$$U_{n\mu} \to V_n U_{n\mu} V^{\dagger}_{n+\hat{\mu}} , \quad q_n \to V_n q_n , \quad \bar{q}_n \to \bar{q}_n V^{\dagger}_n . \quad (5)$$

where V_n is an arbitrary SU(3) matrix defined at each site. Secondly the path integral (3) may be regarded as the partition function of a classical statistical-mechanical system in four dimensions with the 'Hamiltonian' $H=S_g+S_q$, and the matrix element (4) the statistical average in such a system. The usefulness of this connection will become clear shortly.

How do we examine the possibility of confinement in lattice QCD? A useful quantity for this purpose is the potential $V_{q\bar{q}}(r)$ between a static quark and an antiquark at a separation r. A linear potential $V_{q\bar{q}}(r) \sim \sigma r$ will signal confinement. The coefficient σ is called the string tension.

By definition a static quark has an infinite mass and so does not move in the spatial directions. It propagates in time, however, and this temporal propagation is realized by the 'hopping' term $\bar{q}_n \gamma_4 U_{n4} q_{n+\hat{4}}$ in (2). Thus every time the quark hops one lattice spacing in the temporal direction, it leaves behind the gauge variable U_{n4}. Suppose the temporal size of the lattice T is large but finite. For definiteness we shall connect the t=0 and t=T surfaces with the periodic boundary condition. The propagation of a static quark at a spatial site $\vec{n}$ from t=0 to T then generates a gauge field factor $\Omega_{\vec{n}} = \mathrm{tr}(\prod_t U_{(\vec{n},t)4})$ called the Polyakov line [14]. An antiquark at $\vec{m}$ moves in the negative time direction and so generates $\Omega^{\dagger}_{\vec{m}}$. Since the path integral represents the matrix element of $\exp(-TH_{QCD})$ with H_{QCD} the physical Hamiltonian of QCD, one sees that the potential is related to the Ω's by

$$\langle \Omega_{\vec{n}} \Omega^{\dagger}_{\vec{m}} \rangle = \exp(-T \, V_{q\bar{q}}(|\vec{n}-\vec{m}|)) . \quad (6)$$

This equation shows that the potential is determined by the space-time correlation of the gauge field. To understand the properties of the correlation, let us invoke the analogy with statistical mechanics and regard the gauge variable $U_{n\mu}$ as a matrix-valued 'spin' residing on links. If one ignores quarks for the moment, the equations (1) and (3) show that these spins interact through a four-spin coupling and are at an effective temperature $'T'=g^2$. At high temperatures, the spins execute random thermal motions and one expects them to be uncorrelated at large distances. More precisely the spin-spin correlation should decrease exponentially with distance. Thus for strong coupling $g^2 >> 1$ we expect the behavior

$$\langle \Omega_{\vec{n}} \Omega^{\dagger}_{\vec{m}} \rangle \to \exp(-\alpha r), \quad r = |\vec{n}-\vec{m}| \to \infty \quad , \tag{7}$$

with some constant α. Because of the relation (6) this means $V_{q\bar{q}}(r) \sim \sigma r$ for $r \to \infty$, namely confinement. A more explicit calculation using an expansion in $1/g^2$ shows the correctness of this intuitive argument. It also shows that the color flux emanating from quarks is in fact squeezed into a string.

The inclusion of quarks changes this picture slightly. For large r, it becomes energetically more favorable to split the string in the middle by creating a quark-antiquark pair out of the vacuum. Thus the potential will flatten off at large distances in full QCD.

For our purpose the lattice is simply a mathematical device to define QCD. The physical results to be compared with experiment are obtained in the continuum limit $a \to 0$. For this limit to be meaningful the underlying lattice structure should become invisible if the distance is measured in units of some fundamental length scale ξ characterizing QCD. Since the lattice spacing a goes to zero, the length scale ξ/a measured in units of a has to diverge. This is a behavior typical of a second order phase transition. For QCD, there is much evidence that this phase transition occurs at 'T_c'$=g_c^2=0$ and nowhere else. The continuum limit is therefore given by $a\to 0$ and $g^2 \to 0$. The absence of second order phase transitions at $g^2 \neq 0$ means that QCD is in a single phase over the entire range of the coupling constant $0 \leq g^2 \leq \infty$. Thus the confinement property, which we have shown for $g^2 >> 1$, will also hold in the continuum limit $g^2 \to 0$.

Near the critical coupling $g_c^2=0$, physical observables exhibit a particular dependence on g^2 called scaling. For an observable F having the dimension of $(\text{mass})^{\ell}$, this is given by

$$F \sim C \Lambda_L^{\ell} \; , \quad \Lambda_L = \frac{1}{a}(b_0 g^2)^{-b_1/2b_0^2} \exp(-\frac{1}{2b_0 g^2}) \; , \tag{8}$$

where C is a constant and $b_0=(11-2/3N_f)/16\pi^2$ and $b_1=(102-38/3N_f)/(16\pi^2)^2$ with N_f the number of quark flavors. The parameter Λ_L is called the lattice Λ parameter. In the continuum limit, g^2 and a should be varied simultaneously so that Λ_L converges to a finite value. We also note that the factor $\exp(-1/2b_0g^2)$ clearly shows the non-perturbative nature of the dynamics of QCD.

3. Numerical Simulations

Quantitative verification of QCD requires evaluation of physical observables in lattice QCD near the continuum limit. As may be expected from the form of scaling (8), this is not an easy task and analytical methods are not yet available. Instead, most of the recent efforts are concentrated on direct numerical evaluation of the averages defined by (3-4) on a finite but large enough lattice. This is the so-called Monte Carlo simulation[15] based on the method of importance sampling. Let X represent a point of the phase space of field variables and S(X) the value of the action for that point. If one can generate a sequence of points $X_1, X_2, \ldots, X_n$ by an iterative procedure such that the distribution of the points converges to $\exp(-S(X))$ in the limit $n\to\infty$, then the expectation value of an observable $\mathcal{O}$ with respect to the weight $\exp(-S)$ is given by the $n\to\infty$ limit of the average $1/n \sum_{i=1}^{n} \mathcal{O}(x_i)$.

In the context of lattice QCD the Monte Carlo simulation was first used by Creutz and others[9-11] for pure gauge systems. In particular the result of Creutz [10] indicated strongly that the string tension for the quark-antiquark potential exhibits the scaling behavior for weak coupling. This was the first concrete evidence that lattice QCD does have a sensible continuum limit with the confinement property, in agreement with the theoretical arguments of Sec. 2.

More recent applications of Monte Carlo methods aiming at obtaining more precise numerical results have encountered several technical difficulties which have not yet been completely overcome. The first is the large number of degrees of freedom and the associated problem of speed and memory size of computers. To illustrate this point, consider calculating the mass of hadrons numerically. The typical radius of hadrons is about 1 fm and the size of the lattice obviously has to be at least twice this value. If one uses the lattice spacing a=0.1 fm, the number of lattice points per axis is at least 20 which means the number of degrees of freedom of the U's $> 20^4 \times 4 \times 3 \times 3 \times 2$ real numbers = 88 megabytes with 64 bit precision. If one doubles the lattice size the memory requirement increases $2^4=16$ fold to 1.37 gigabytes. This many numbers represents a single point in the phase space of the U's. In an update of the entire lattice (i.e., change the values of all U's to generate a new point from an old one in the phase space), one obviously has to carry out many more arithmetic operations. Since several thousand updates or more are normally necessary for reasonable statistics, the speed and memory of computers have to be in the range of several hundred megaflops($=10^6$ floating point operations per second) and megabytes. The second problem, which is more complex and leads to a demand of much higher speed, is the Grassmann nature of the quark variables in the path integral (3-4). These are anti-commuting numbers which cannot be directly manipulated in normal computers. Fortunately the quark action generally takes the form $S_q = \Sigma \bar{q} D(U) q$ with D(U) the lattice Dirac operator, and the path integral can be carried out with the result detD(U). If the operator D(U) is positive, it is also possible to rewrite the determinant as $\int \pi dY dY^{\dagger} \exp(-\Sigma Y^{\dagger} D(U)^{-1} Y)$ with Y a complex field defined on sites. These expressions only use ordinary numbers and hence are amenable to the standard Monte Carlo methods. However, they both involve a non-local quantity (detD or D^{-1}) which is extremely time-consuming to evaluate. For example, consider solving an equation Dx=Y to obtain $x=D^{-1}Y$. In the existing algorithms for solving such an equation, a typical operation is the multiplication $p\to Dp$ with p a vector in site, color and Dirac indices. If one uses the 20^4 lattice, the dimension of D is $20^4 \times 4$(for Dirac)$\times 3$(for color)$=1.92\times 10^6$ and the number of arithmetic operations for the above multiplication is over 370×10^6. Since this multiplication has to be repeated at least several dozen times to find $D^{-1}Y$ and since it has to be calculated for each update, inclusion of quarks requires a computer which is several dozen times faster than the one necessary for the pure gauge system. These demands are extremely severe and are only beginning to be met by the most recent supercomputers.

The Monte Carlo method for the pure gauge system is now quite standard [10,11]. A major concern in recent years has been to develop an efficient algorithm for incorporating quarks in the simulation [16-19]. One of the proposals which appears to be quite promising is the Langevin method [18] based on the stochastic quantization [20] and we shall describe it briefly. Let ϕ_i be the field variable with i the lattice and other indices and $S(\phi)$ the action of the system. In the stochastic quantization, one introduces a fictitious time τ and assumes that the τ dependence of ϕ_i is given by the stochastic equation

$$\frac{d}{d\tau}\phi_i(\tau) = -\frac{\delta S(\phi(\tau))}{\delta\phi_i(\tau)} + \eta_i(\tau) , \tag{9}$$

where $\eta_i(\tau)$ is a white noise in τ. Using the Fokker-Planck equation one can show [20] that the distribution of the trajectory of the solution to (9) converges to $\exp(-S(\phi))$ in the limit $\tau\to\infty$. The expectation value of an observable can therefore be evaluated by taking the average with respect to τ. The numerical implementation of the stochastic equation requires a discretization of the fictitious time τ. Actually this introduces a systematic error since the distribution of configurations will deviate from $\exp(-S)$ for a finite discretization size $\Delta\tau$. However the analyses of the Fokker-Planck equation corresponding to the discretized Langevin equation allow analytical studies of such errors.

To apply the Langevin method to QCD [18] one uses the effective action $S_{eff} = S_g(U) + \Sigma\, Y^{\dagger} D(U)^{-1} Y$ taking the gauge field $U_{n\mu}$ and the 'pseudo-fermionic' field Y_n as the basic variables. The Langevin method has several nice features which do not always apply to other proposals for incorporating quarks. The distribution of field configurations according to $\exp(-S_{eff})$ in the limit $\Delta\tau\to 0$ is guaranteed by the stochastic nature of the Langevin equation and the systematic error due to a finite $\Delta\tau$ is in principle controllable. The cost of including quarks is reduced to solving a linear equation of the form $D(U)x=Y$ once per update.

4. Hadron Mass Calculations

The lattice QCD and numerical simulation techniques provide a method for calculating the observables of strong interactions. In particular they enable the calculation of the mass spectrum and other static properties of hadrons. The usual method for calculating masses utilizes Green's functions. If $O_H(t)$ denotes an operator that creates or annihilates a hadron H at time t with zero spatial momentum (for example $O \sim \bar{q}\gamma_5 q$ for pseudo-scalar mesons and $O \sim qqq$ for baryons), the Green's function for O_H behaves as

$$<O_H^\dagger(t)\ O_H(0)> = \sum_{m_i < m_0} A_i e^{-m_i t} + \int_{m_0}^{\infty} dm\ \rho(m) e^{-mt} , \tag{10}$$

where the second term represents the continuum contributions. Thus the hadron mass can be calculated from the decay rate of the corresponding Green's function with respect to time t.

The lattice approach requires the masses of some number of hadrons to fix the mass scale. If there are N_f quark flavors a hadron mass m_H calculated on a lattice has the functional dependence given by $m_H = 1/a\ f_H(m_1 a, \ldots, m_{N_f} a, g^2)$ with $m_i, i=1,\ldots,N_f$, the quark masses. As we noted in Sec. 2, the variation of the lattice spacing a and the coupling constant g^2 are related through the Λ parameter. Thus there are N_f+1 free parameters $m_1,\ldots,m_{N_f}$ and a (or Λ_L), and an equal number of hadron masses has to be used as inputs to fix their values. For example if one considers hadrons made up of u and d quarks and assumes an exact isospin symmetry $m_u = m_d$, one needs two masses, e.g., m_π=140MeV and m_ρ=770MeV. The masses of the other hadrons, e.g., nucleon and delta, are predictions of QCD to be compared with experiment.

The first attempt at the mass spectrum calculation was made by Hamber and Parisi and by others in 1981[21-23]. Using the π and ρ masses as inputs Hamber and Parisi obtained m_N=950±100MeV and m_Δ=1300±100MeV as compared with the experimental values 940MeV and 1230MeV. Actually these results should be viewed with some care for several reasons. First of all the calculation was done on a $6^3\times 10$ lattice at the coupling constant $\beta = 6/g^2 = 1.0$. The lattice spacing turned out to be 0.18 fm and so the spatial lattice size is 6×0.18=1.08 fm. This may be barely large enough to contain a pion (charge radius 0.57 fm) but clearly not for a nucleon (proton charge radius is 0.83 fm). Secondly, because the inclusion of quarks is extremely

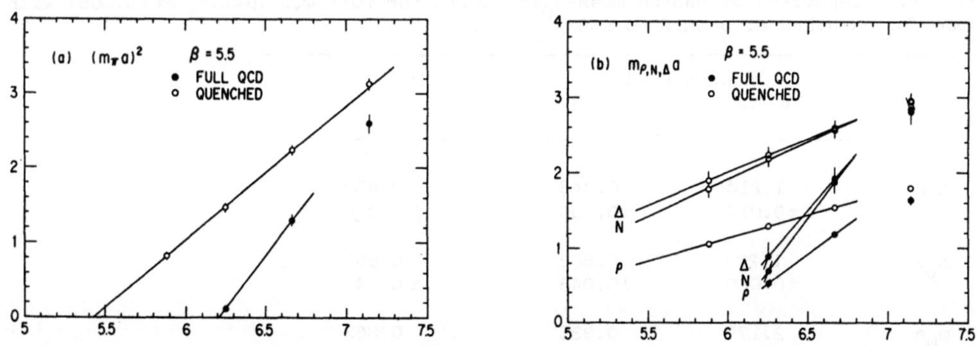

Fig. 1. Comparison of (a) pion mass squared and (b) rho, nucleon and delta masses in lattice units calculated in full QCD (N_f=2) and quenched QCD with Wilson quark action on a $9^3\times 18$ lattice. The value of coupling constant is $\beta = 6/g^2 = 5.5$. The quark mass may be defined as $m_q a = (1/K - 1/K_c)/2$ where K_c denotes the value of K at which the pion mass vanishes.

time-consuming, Hamber and Parisi, and also the others [22,23], used the 'quenched approximation' in which the quark determinant detD(U) is ignored in gauge updates. This means neglecting the effects of $\bar{q}q$ pair creation and annihilation whose validity is not known.

In spite of these problems the results of the initial attempts were certainly encouraging and many calculations, albeit within the quenched approximation, have subsequently been carried out on an increasingly larger lattice and with better statistics. To quote the most recent result, Ito, Iwasaki and Yoshie[24] used a $16^3\times48$ lattice and paying particular attention to the finite lattice size and finite lattice spacing effects obtained the estimate m_N=1080±80MeV and m_Δ=1370 ±120MeV. Although the scaling behavior (8) has not been checked, this result indicates that the quenched approximation predicts the masses in agreement with experiment to within 10 to 15%.

The full QCD calculation of the mass spectrum without relying on the quenched approximation did not start till last year [25,26] because of the technical difficulties described in Sec. 3. Using the Langevin method[18], Fukugita, Oyanagi and myself[25] carried out a simulation on a $9^3\times18$ lattice for N_f=2 flavors with the Wilson quark action [13]. The lattice spacing turned out to be 0.12 fm indicating that our lattice may not be large enough for a realistic estimate of baryon masses. The basic features of the effects of quarks have clearly emerged, however. The full QCD value of masses in lattice units $m_H a$ is substantially decreased from the quenched value at the same coupling constant $\beta=6/g^2=5.5$ (see fig. 1 and compare the first and last column of table 1). This is expected to a certain extent since the vacuum polarization of quarks screens the color charge and so decreases the effective value of g^2. An important result is that this difference disappears within our statistics for ground state hadrons if one shifts

Table 1. Comparison of hadron masses at K=0.15 for full QCD (β=5.5, $\Delta\tau$=0.005) with those for quenched at a shifted $\beta=6/g^2$=5.79.

	quenched QCD		full QCD
	β=5.5	5.79	β=5.5
$m_\pi a$	1.214 ±0.017	0.448 ±0.036	0.458 ± 0.026
$m_\rho a$	1.293 ±0.020	0.602 ±0.048	0.602 ± 0.041
$m_N a$	2.176 ±0.099	0.935 ±0.201	0.863 ± 0.117
$m_\Delta a$	2.229 ±0.107	1.145 ±0.217	0.994 ± 0.176

the value of the gauge coupling used in the quenched calculation by an appropriate amount (see table 1 for an example). The value of g^2 is not a physical quantity and this agreement means that, at the level of accuracy of our data($\sim$10-20%), the hadron masses in physical units should coincide between the quenched and full QCD. The physical reason for this behavior is that the pure gauge theory potential for quarks at the shifted coupling approximately coincides with the full QCD potential at short and intermediate distances relevant for ground state hadrons. These features of full QCD are the underlying reason for the closeness of the quenched mass spectrum to the experimental one noted above.

There are many problems that remain to be elucidated in the full QCD simulation. The most interesting is the simulation at small enough quark masses for which decays of hadrons such as $\rho \rightarrow \pi\pi$ and $\Delta \rightarrow N\pi$ are allowed. For a realistic estimate it is also important to use a larger lattice to avoid finite size effects and to check the scaling behavior.

5. QCD at High Temperatures

The discussions so far have tacitly assumed zero temperature and vanishing baryon number density. There exist situations such as the evolution of the early Universe, the possibility of quark stars and high energy heavy ion collisions in which the temperature and/or the baryon number density is expected to become extremely high. It has been speculated for some time[27] that the hadronic matter might undergo a change of phase in such unusual environments. With the advent of QCD it has become possible to examine this problem quantitatively.

Let us first consider the situation at finite temperatures. It has been argued that the confinement property will be lost at sufficiently high temperatures [14]. A naive argument runs as follows. At zero temperature the static $\bar{q}q$ potential increases linearly with distance due to the formation of the color string. At non-zero temperatures, this string will fluctuate thermally. The free energy of the string per unit length will therefore be lowered and eventually might vanish at a certain temperature. If this happens, the quarks are no longer bound by a linear potential and will become liberated.

Another property of strong interactions that may change at high temperatures is that related to chiral symmetry. This is the symmetry under rotation of quark flavors with the positive and negative chiral components ($\gamma_5=\pm 1$) rotated in the opposite direction. At zero temperature, this symmetry is spontaneously broken by the condensation of $\bar{q}q$ pairs of opposite chirality and the pion is the Nambu-Goldstone boson of this breaking. At sufficiently high temperatures, however, the thermal energy will break up the $\bar{q}q$ condensate. This will result in a restoration of the chiral symmetry.

These arguments suggest that the dynamics of QCD changes drastically with increasing temperature, possibly through phase transitions at a several hundred MeV. The early perturbative studies[28] indicated that the high-temperature phase of QCD could be characterized as a plasma of quarks and gluons. The perturbative methods are not powerful enough, however, for quantitative analyses of the phase transition itself. For this purpose, the lattice formulation and the numerical simulations have again proved to be a valuable tool. To simulate QCD at a finite temperature, one uses a lattice whose spatial extent is large compared with its temporal size N_t. The temperature T is then equal to

$$T = \frac{1}{N_t a} \quad . \tag{11}$$

It may seem that the temperature cannot be varied continuously on a lattice. We recall, however, that the lattice spacing a and the coupling constant g^2 simultaneously decrease in the continuum limit. Therefore one can continuously increase the temperature by decreasing the coupling constant.

The nature of phase transitions can be analyzed by the behavior of the order parameter. For the confining property the order parameter is given by the Polyakov line $\Omega_{\vec{n}}=\mathrm{tr}(\prod_t U_{(\vec{n},t)4})$ which we have already encountered in Sec. 2 and which represents the interaction of a static quark at the spatial site $\vec{n}$ with the gauge field. At a non-zero temperature, the average value of Ω is related to the free energy F_q of a static quark by

$$\langle\Omega_{\vec{n}}\rangle = \exp(-F_q/T) \quad . \tag{12}$$

Thus $\langle\Omega_{\vec{n}}\rangle=0$ means confinement ($F_q=\infty$) whereas this property will be lost if $\langle\Omega_{\vec{n}}\rangle$ becomes non-zero [14]. Strictly speaking, the latter statement is correct only if quarks are absent because otherwise the color charge of the static quark will always be screened by the creation of $\bar{q}q$ pairs, leading to a finite value of F_q. It turned out in practice, however, that $\langle\Omega\rangle$ changes from a small to a finite value over a narrow range of temperature and in this sense provides a useful quantity for phase structure analyses. For the chiral symmetry the order parameter is the $\bar{q}q$ operator. In the limit of vanishing quark mass $m_q=0$, the chiral symmetry is spontaneously broken if $\langle\bar{q}q\rangle \neq 0$ and is restored when $\langle\bar{q}q\rangle = 0$.

We now summarize the current picture of the phase structure of QCD as a function of temperature T and quark mass m_q. In the pure gauge limit $m_q=\infty$ it is well established [29,30] that there exists a first order deconfining phase transition. In other words, the physical observables jump across the transition. Perturbative analyses in $1/m_q$ [31] show that this first order transition should weaken as m_q decreases from ∞. This has been confirmed by the numerical studies [32,33]. A priori the deconfining transition can disappear below some value of m_q, while the chiral transition at $m_q=0$, possibly extended into a finite range of m_q, may appear at a completely different temperature. For quarks in the fundamental representation of SU(3), however, the abrupt changes of $\langle\Omega\rangle$ and $\langle\bar{q}q\rangle$ always occur simultaneously and the deconfining transition smoothly continues on to the chiral transition with decreasing m_q. At intermediate values of m_q, the transition might turn into a sharp but continuous crossover. (See ref. 32-35 for relevant numerical studies.) An interesting feature[32,35] (which has gained a consensus for the $N_f=4$ case [36]) is that the transition becomes a first order phase transition for sufficiently small m_q. We note for comparison that a theoretical analysis[37] predicted that the chiral transition at $m_q=0$ is first order for $N_f=3$ or more flavors.

The qualitative features summarized above are illustrated in figs. 2-3 which are taken from our Langevin simulation[33] on a $8^3\times4$ lattice with $N_f=4$. One sees in fig. 2 that the hysteresis in thermal cycles, indicative of a first order transition, disappears and then reappears as the quark mass is decreased. Fig. 3 shows evidence for a first order chiral transition at small m_q; the high and low temperature phases coexist as the system evolves in Langevin time τ.

The full QCD finite temperature simulations done so far used fairly small lattices and mainly addressed qualitative questions such as the order of the transition. An important problem which awaits future simulations on a larger

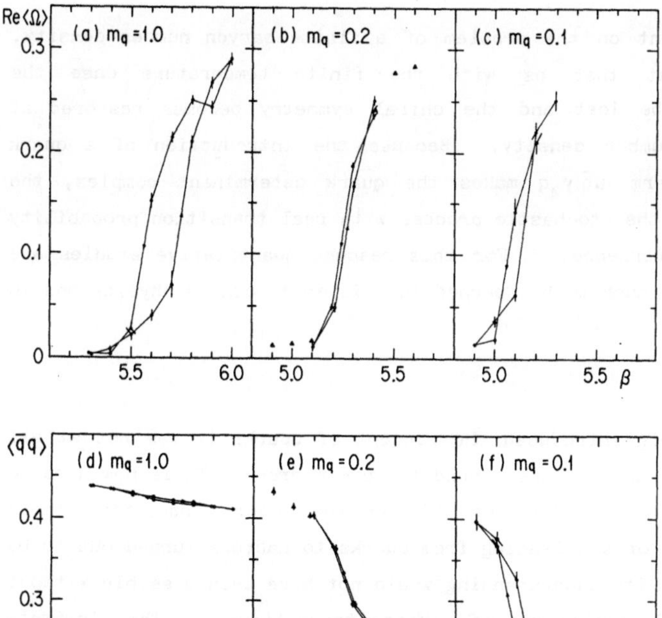

Fig. 2. Behavior of (a-c) the Polyakov line and (d-f) the chiral order parameter in thermal cycles on a $8^4\times4$ lattice with 4 flavors. The value of quark mass m_q is in lattice units. The temperature effectively increases toward larger value of the coupling constant $\beta=6/g^2$.

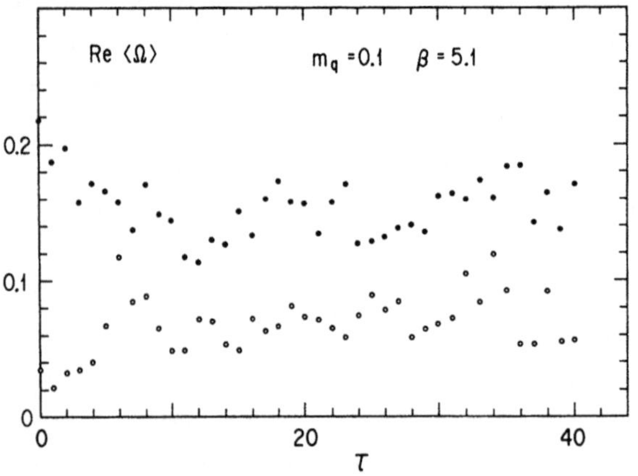

Fig. 3. Langevin time evolution of the Polyakov line in two simulations at $\beta=6/g^2$ $=5.1$ and $m_q=0.1$ in lattice units, one starting from the high-temperature phase (solid circle) and the other from the low-temperature phase(open circle).

lattice and at realistic quark masses is the determination of values of observables characterizing the transition, such as the critical temperature and the internal energy density.

We shall briefly comment on the problem of a finite baryon number density. Intuitive arguments suggest that as with the finite temperature case the confinement property will be lost and the chiral symmetry becomes restored at sufficiently high baryon number density. Because the introduction of a quark chemical potential by a term $\mu\bar{q}\gamma_4 q$ makes the quark determinant complex, the simulation methods based on the stochastic process with real transition probability loses the guarantee of convergence. For this reason, quantitative studies are still scarce. There remains much to be learned in this area both in physics and in techniques.

6. Summary and Outlook

When Gell-Mann and Zweig introduced the concept of quarks it was not at all clear whether they were real nor how they bound to form hadrons. It is now hard to doubt their reality and the validity of QCD as the theory describing their interactions. The dynamics of QCD leading from quarks to hadrons turned out to be a complicated one. In fact its understanding would not have been possible without help from the statistical mechanics of phase transitions. The intimate relationship between quantum field theory and statistical mechanics is not restricted to QCD and has been known for some time [38]. It has yielded particularly fruitful results in the case of QCD and this has been a source of major stimulus for recent non-perturbative studies of other field theories.

The quantitative verification of QCD has not yet reached the final stage. At the moment, because of the complexity of its dynamics, the numerical simulations are the most powerful method for extracting the predictions of QCD. As we have mentioned several times, these simulations require truly powerful computers. To quote an example, our Langevin simulation of the full QCD hadrom mass spectrum[25] used 67 megabytes of memory and about 1000 hours of CPU time on HITAC S810/10 which has the peak speed of 315 megaflops. This experience points toward 10 gigaflops or more for speed and at least a few gigabytes for memory to carry out full QCD simulations with a realistic predictive power on a sufficiently large lattice. This marks the lattice QCD simulations as one of the major problems in computational physics. Fortunately the progress of computer technology is remarkably fast and computers with such a capacity might become available shortly. There is much hope that, half a century after its beginning, the theory of strong interactions in the form of QCD will soon reach the stage where we will be able to carry out detailed calculations of the strong interaction observables from first principles.

References

1. H. Yukawa, Proc. Phys.-Math. Soc.(Japan), 17, 48 (1935).
2. C.M.C. Lattes et al., Nature 159, 694 (1947).
3. Review of Particle Properties, Phys. Lett. 170B, 1 (1986).
4. M. Gell-Mann, Phys. Lett. 8, 214 (1964); G. Zweig, CERN preprint 8182/TH 401 (1964).
5. C.N. Yang and R.L. Mills, Phys. Rev. 96, 1605 (1954); R. Utiyama, Phys. Rev. 101, 1597 (1956).
6. W. Bardeen, H. Fritsch and M. Gell-Mann, in Scale and Conformal Invariance in Hadron Physics, ed. R. Gatto (Wiley, New York, 1973), p.139.
7. D.J. Gross and F. Wilczek, Phys. Rev. Lett. 30, 1343 (1973); H.D. Politzer, Phys. Rev. Lett. 30, 1346 (1973).
8. K.G. Wilson, Phys. Rev. D10, 2445 (1974); A.M. Polyakov, unpublished.
9. M. Creutz, L. Jacobs and C. Rebbi, Phys. Rev. Lett. 42, 1390 (1979).
10. M. Creutz, Phys. Rev. D21, 2308 (1980); Phys. Rev. Lett. 45, 313 (1980).
11. K.G. Wilson, in Recent Developments in Gauge Theories, eds. G. t'Hooft et al., Cargese 1979 (Plenum, New York, 1980), p.363.
12. H.B. Nielsen and M. Ninomiya, Nucl. Phys. B185, 20 (1981).
13. K.G. Wilson, in New Phenomena in Subnuclear Physics, ed. A. Zichichi (Plenum, New York, 1979, p.69); L. Susskind, Phys. Rev. D16, 3031 (1977).
14. A.M. Polyakov, Phys. Lett. 72B, 477 (1978); L. Susskind, Phys. Rev. D20, 2610 (1979).
15. N. Metropolis et al., J. Chem. Phys. 21, 1087 (1953).
16. F. Fucito et al., Nucl. Phys. B180 [FS2]; 369 (1981).
17. J. Polonyi and H.W. Wyld, Phys. Rev. Lett. 51, 2257 (1983).
18. A. Ukawa and M. Fukugita, Phys. Rev. Lett. 55, 1854 (1985); G.G. Batrouni et al., Phys. Rev. D32, 2736 (1985).
19. S. Duane, Nucl. Phys. B257 [FS14], 652 (1985); S. Duane and J.B. Kogut, Nucl. Phys. B275 [FS17], 398 (1986).
20. G. Parisi and Y.-S. Wu, Sci. Sin. 24, 483 (1981).
21. H. Hamber and G. Parisi, Phys. Rev. Lett. 47, 1792 (1981).
22. D. Weingarten, Phys. Lett. 109B, 57 (1982).
23. A. Hasenfratz et al., Phys. Lett. 110B, 289 (1982).
24. S. Ito , Y. Iwasaki and T. Yoshie, Phys. Lett. 183B, 351 (1987). See also D. Barkai et al., Phys. Lett. 156B, 385 (1986).
25. M. Fukugita, Y. Oyanagi and A. Ukawa, Phys. Rev. Lett. 57, 953 (1986).
26. F. Fucito et al., Phys. Lett. 172B, 235 (1986).
27. R. Hagedorn, Nuovo Cim. Suppl. 3, 147 (1965); N. Cabbibo and G. Parisi, Phys. Lett. 59B, 67 (1975).
28. D.J. Gross et al., Rev. Mod. Phys. 53, 43 (1981) and references cited therein.

29. B. Svetitsky and L. Yaffe, Nucl. Phys. B210 [FS6], 423 (1982).
30. K. Kajantie et al., Z. Phys. C9, 253 (1981); S.A. Gottlieb et al., Phys. Rev. Lett. 55, 1958 (1985); N. Christ et al., Phys. Rev. Lett. 56, 111 (1986).
31. T. Banks and A. Ukawa, Nucl. Phys. B225 [FS9], 145 (1983); T. De Grand and C. De Tar, Nucl. Phys. B225 [FS9], 590 (1983).
32. P. Hasenfratz et al., Phys. Lett. 133B, 221 (1983); T. Celik et al., Phys. Lett. 133B, 427 (1983).
33. M. Fukugita and A. Ukawa, Phys. Rev. Lett., 57, 503 (1986).
34. J. Polonyi et al., Phys. Rev. Lett. 53, 644 (1984); R.V. Gavai, Nucl. Phys. B269, 530 (1986); J.B. Kogut and D.K. Sinclair, Nucl. Phys. B280 [FS18], 625 (1987).
35. F. Fucito and S. Solomon, Phys. Rev. Lett. 55, 2541 (1985) and references cited therein. See also R.V. Gavai and F. Karsch, Nucl. Phys. B261, 273 (1985).
36. R. Gupta et al., Phys. Rev. Lett. 57, 2621 (1986); J.B. Kogut et al., Phys. Rev. Lett. 58, 751 (1987).
37. R. Pisarski and F. Wilczek, Phys. Rev. D29, 338 (1984).
38. K. Symanzik, J. Math. Phys. 7, 510 (1966); K.G. Wilson and J.B. Kogut, Phys. Rep. C12, 75 (1974).

The Very Early Universe

M. Yoshimura
National Laboratory for High Energy Physics (KEK),
Tsukuba, Ibaraki 305, Japan

In this short talk I would like to explain some exciting new developments taking place at the interface between particle physics and cosmology. This is a subject well suited to this kind of conference that consists of the audience from many different disciplines, but demands that the speaker explains in simple terms many concepts unfamiliar to most of the audience. I shall do my best, although success is not guaranteed.

I. Big Bang Model

Our universe is a site of many different forms of astronomical objects: their sizes range from 10^{10} cm for stars, to 10^{23} cm for galaxies, up to 10^{28} cm for the observable part of the universe. Correspondingly, their masses lie between 10^{33} gr for stars, 10^{44} gr for galaxies, and 10^{54} gr or more for the observable universe. These values are subject to variations and considerable measurement uncertainties, and should be taken only as a rough idea. It is remarkable that a reliable measurement of the distance to the nearest Andromeda nebula was established only fairly recently in 1952. Our subject in this talk, cosmology, deals with the present universe on a much larger scale: in the first approximation galaxies are uniformly distributed in this scale.

Although the universe is static in the time scale of human beings, in reality it evolves. This picture emerged from two crucial observations. First, Hubble established a relation between the distance (a) to far galaxy and its recession velocity ($\dot{a}$): $\dot{a} \propto a$. The Hubble law can be understood by balancing kinetic energy against gravity in a uniform and isotropic universe:

$$\frac{1}{2}\dot{a}^2 \sim G\frac{4\pi}{3}a^3\rho/a, \quad \text{and therefore} \quad \dot{a}/a \sim (8\pi G\rho/3)^{1/2}. \qquad (1)$$

An important outcome of the Hubble expansion is that the inverse of the Hubble constant $(\dot{a}/a)^{-1}$ gives a time scale, which turns out to be $\sim$15 giga years (Gy $\equiv 10^9$y). This is longer than the age of the sun (4.6 Gy), but is comparable to the age of globular clusters in our galaxy.

More dramatic evidence for the expanding universe was given by the discovery of the microwave background radiation at 3K. It became clear in later measurements at various wavelengths that the spectrum of this radiation follows the Planck distribution. The most natural interpretation of this radiation is that it is a relic of a hot initial universe, the wavelength (λ) being red shifted by the expansion according to the law, $\lambda \propto a(t)$. If one assumes that the overall size of the universe changes as a power law, $a(t) \propto t^{\gamma}$, the Hubble parameter $H = \dot{a}/a = \gamma/t$. Hence H^{-1} is related to the age of the universe t. It is estimated that the size has increased by roughly a factor of 10^3 since last scattering of the microwave (then optical) radiation, which occurred at the moment of photon decoupling from matter.

Although one can learn much from Newtonian gravity, one has to rely on Einstein's general relativity if one wants to discuss the universe on the largest scale. In addition to the fundamental law of gravity, one assumes in the big bang model the cosmological principle: the universe is homogeneous (every point in space is on the same footing) and isotropic (there is no preferred direction in space). This is of course a hypothesis that has to be checked by observation. The cosmological principle simplifies the allowed form of the space-time metric which contains two adjustable parameters:

$$ds^2 = dt^2 - a^2(t)\left((1-kr^2)^{-1}dr^2 + r^2(d\theta^2 + \sin^2\theta d\phi^2)\right) . \qquad (2)$$

The scale factor a(t) describes how the overall size of the universe changes as a function of the cosmic time t. The parameter k can take any one of the three values; ±1, 0, and distinguishes the global topology of space: for $k = 1$, the space is 3-sphere (three-dimensional sphere that can be embedded in a flat four-dimensional Euclidean space) and for $k = -1$ and $k = 0$, the space is unlimited (Fig. 1). The spatial curvature of the metric (2) is given by $R^{(3)} = 6ka^{-2}$, and only when $k = 0$, is the space flat.

With the Robertson-Walker metric (2), the Einstein equation takes a simple form of

$$(\dot{a}/a)^2 + ka^{-2} = \frac{8\pi}{3} G \rho. \qquad (3)$$

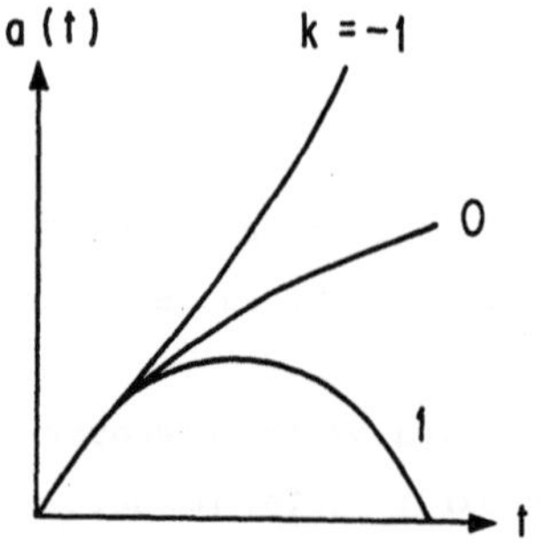

Fig. 1 Scale factor

Compared to the Newtonian equation (1), the curvature term of $k\,a^{-2}$ is a modification due to general relativity. Only when $k > 1$ ($k = 1$), can $\dot{a}$ vanish, giving a possibility of the contracting phase.

The other Einstein equation is equivalent to the Bianchi identity, $R^{\mu}{}_{\nu;\mu} = 0$ obeyed by the Ricci curvature tensor, which then implies a local conservation of the energy-momentum tensor; $T^{\mu}{}_{\nu;\mu} = 0$. With (2), it is given by

$$(\rho a^3)^{\cdot} = -p\,(a^3)^{\cdot} \quad . \tag{4}$$

Two differential equations, (3) and (4), for 3 variables, a, ρ and p, become integrable when they are supplemented by an equation of state, $p(\rho)$. Two cases are of prime importance; radiation-dominated (RD) epoch and matter-dominated (MD) epoch.

In the MD epoch which is applied to the present universe, the pressure is very small compared to the energy density, $p \ll \rho$, and the energy density ρ is dominated by nonrelativistic particles such as nucleons; $\rho = mn \propto a^{-3}$. In the flat universe of $k = 0$, (3) yields

$$a(t) \propto t^{2/3} \quad (k = 0,\ \mathrm{MD}) \ . \tag{5}$$

For $k = \pm 1$, the solution takes a more complicated form;

$$k = 1: \quad a = c\,\sin^2\theta \ , \quad t = c(\theta - \tfrac{1}{2}\sin 2\theta); \tag{6a}$$

$$k = -1: \quad a = c\,\sinh^2\theta \ , \quad t = c(\tfrac{1}{2}\sinh 2\theta - \theta) \ . \tag{6b}$$

Effect of the curvature shows up at moderately large times. For early times with $|\theta| \ll 1$, both of (6a) and (6b) yield the same behavior as (5). The present universe is still in this early phase: one can not observationally determine the curvature effect. An important measure of the curvature term is given by the Ω-parameter,

$$\Omega \equiv 8\pi G\rho/(3H^2) \ , \tag{7}$$

which is equal to $1 + k\,\dot{a}^{-2}$ by the Einstein equation. The universe is open ($k \leqslant 0$) or closed ($k > 0$), depending on whether $\Omega \leqslant 1$ or $\Omega > 1$. In principle, one should be able to check a consistency relation between the Hubble parameter H, the age t, and Ω, given by

$$Ht = 2/3 \quad (\Omega = 1) \ , \qquad Ht = f(\Omega) \ ;$$

$$f(x) = \begin{cases} (1-x)^{-1} - \frac{1}{2} x(1-x)^{-3/2} \ln\left(x^{-1}(2-x+2\sqrt{1-x})\right), & (x<1) \\ -(x-1)^{-1} + x(x-1)^{-3/2} \text{ arc cos } x^{-1/2}. & (x>1) \end{cases}$$

In practice, our observations are far from being able to verify this relation.

On the other hand, in the RD epoch the universe is dominated by relativistic particles such as the photon. The equation of state is determined by the tracelessness of the energy-momentum tensor, $T^{\mu}{}_{\mu} = 0$, which means that there is no mass scale associated with massless particles. Thus, $p = \rho/3 \propto a^{-4}$ from (4). The curvature term in the Einstein equation (3) can be neglected because its importance relative to the energy density term decreases as $t \to 0$. Hence $\dot{a}/a \propto \sqrt{\rho} \propto a^{-2}$, and

$$a(t) \propto t^{1/2} \text{ (RD)}. \tag{8}$$

The radiation dominated universe is practically made of various almost massless particles in thermal equilibrium. The energy density is then given by

$$\rho = \frac{\pi^2}{30} N_F T^4 . \tag{9}$$

N_F is an effective degree of freedom, counted 1 for a boson species and 7/8 for a fermion species, in thermal equilibrium whose masses are much less than the temperature T (I always use a unit such that $\hbar = c = k_B = 1$). A convenient time-temperature relation then follows;

$$t \sim 2 \times 10^{-6} \text{ sec}/\{\sqrt{N_F}\cdot(T/1 \text{ GeV})^2\}. \tag{10}$$

For instance, light nuclei, as will be explained shortly, are formed at $T \sim 0.1$ MeV and $N_F \sim 43/4$, hence at $t \sim 10^2$ sec.

Exploration of the early universe amounts to elucidation of the microscopic world at high temperatures. The first major event, dating back from the present time to early times, is characterized by plasma recombination, in which radiation decouples from matter when neutral atoms (mainly hydrogen) are formed. This is supposed to have occurred at $T \sim 3000$ K. Since this time photons have propagated freely in space, their wavelength being red shifted by $\sim 10^3$, to become 3K radiation at present. The next major event is formation of light elements (^{4}He, D, ^{3}He, ^{7}Li). This era starts from nucleons (p and n) in thermal equilibrium caused by weak interaction such as $p + e \leftrightarrow n + \nu_e$. The equilibrium abundance $N_n/N_p = \exp(-\Delta m/T)$, with Δm the n-p mass difference, is established until the temperature falls down to $\sim$1 MeV. Nothing much happens after that, except that the abundance of neutrons diminishes due to weak processes including the free neutron decay (its half-life is ∿10 minutes), until accumulation of deuterium rapidly grows at T ∿ 0.1 MeV. This temperature is determined by asking when the photodestruction of deuteron ($D + \gamma \to p + n$) is blocked. By a series of nuclear reactions, ^{4}He is mainly formed, beyond which heavier elements are difficult to process due to the mass gap at A = 5 and 8. The mass fraction of ^{4}He thus formed is fairly insensitive to the nucleon density and initial conditions. Its calculated value of ∿25% agrees well with observation [1]. It is very difficult to produce this large abundance in a stellar interior. The helium abundance can thus be regarded as a major success of the big bang model.

Actually, one can use this quantity to limit many hypothetical theories. The best example is a bound on the number of neutrino species. From high energy experiments it is almost certain that there exist at least three neutrino species; ν_e, ν_μ and ν_τ. The number of neutrino species contributes to the energy density via the factor N_F of (9) (although the neutrino decouples from the rest at ∿1 MeV, their energy density contributes to the expansion time scale, roughly according to (10)). One additional neutrino species gives a faster expansion, hence less time is available for the $n \to p$ conversion, leading to a larger ^{4}He fraction by ∿1.3% [2]. By taking a safe upper bound for the observed ^{4}He fraction, one can establish the allowed number of neutrino species: $N_\nu \lesssim 4$ [2]. Only one more species is allowed, beyond those already known.

Although the big bang model is remarkably successful in presenting a coherent picture, it has also left many unsolved problems. Some of these problems have been used in recent developments as a clue to go beyond the standard big bang model. First of all, the universe appears asymmetrical between matter and antimatter (baryon asymmetry): there is no observational data that suggest presence of bulk antimatter in our cosmos. Although we do not have any direct evidence that distant galaxies are made of matter and not antimatter, the evolutionary picture presents many troubles with a symmetric cosmology in which an equal amount of matter and antimatter exists globally. Among others, the symmetric cosmology [3] yields too small a leftover matter (and antimatter), a factor of $\sim 10^8$ smaller than the observed baryon to photon ratio of $\sim 10^{-10}$. This comes about unless one separates matter from antimatter on a large scale (at least larger than the size of clusters of galaxies) before the temperature falls below ∿20 MeV. In the asymmetric cosmology the present baryon to photon ratio implies the same amount of baryon asymmetry prior to the epoch of nucleon-antinucleon annihilation at 20 MeV. Namely, one excess of baryon over 10^{10} pairs of baryons and antibaryons leads to the present matter content of our universe. This fact needs a definite scientific explanation in my opinion.

The second fundamental problem of the standard big bang model has to do with homogeneity, and is called horizon problem. Except the dipole fluctuation of the

3 K radiation of $\sim 10^{-3}$ in amplitude [4], which is easily interpreted as a local motion of the solar system, no temperature fluctuation has been observed, the upper limit being $(\delta T/T) < 2 \times 10^{-5}$ at a small angular scale of 4.5' [5]. Upon a closer examination this remarkable homogeneity is a mystery. Homogeneity is believed to have occurred by causal processes which can propagate at fastest with light velocity. Thus, homogeneous regions are roughly limited by the particle horizon r_H whose definition is given by the proper distance the light can travel since the beginning of the big bang:

$$r_H \equiv a(t)\int_0^t dt' a^{-1}(t') \ . \tag{11}$$

With a fractional power law of $a(t) \propto t^{\gamma}$ such as (5) and (8), this horizon is of the order ct: more precisely,

$$r_H = (1-\gamma)^{-1} t, \quad \gamma = 2/3 \text{ (MD) or } 1/2 \text{ (RD)}. \tag{12}$$

The surface of last scattering at $T \sim 3000$ K is then divided by a number of causally disconnected regions. This number is estimated as follows: the horizon distance of $\sim t_d$ at decoupling is stretched by a factor of $\sim (t_o/t_d)^{2/3}$ at present, while the surface of last scattering lies at $\sim t_o$ (Fig. 2), thus giving a crude number of causal regions,

$$\left(\frac{t_o}{t_d \cdot (t_o/t_d)^{2/3}}\right)^2 = \left(\frac{t_o}{t_d}\right)^{2/3} = \frac{T_d}{T_o} \sim 10^3 .$$

In retrospect, one is led to wonder why the uniform 3K background exists at all.

The third fundamental problem is the flatness-oldness problem. At no instant of the past has the curvature term of $\sim k\,a^{-2}$ been important. Indeed, the ratio of the curvature to kinetic terms becomes smaller as one goes back in time:

$$(k/a^2)/(\dot{a}/a)^2 = k/\dot{a}^2 \propto t^{2(1-\gamma)} \ . \tag{13}$$

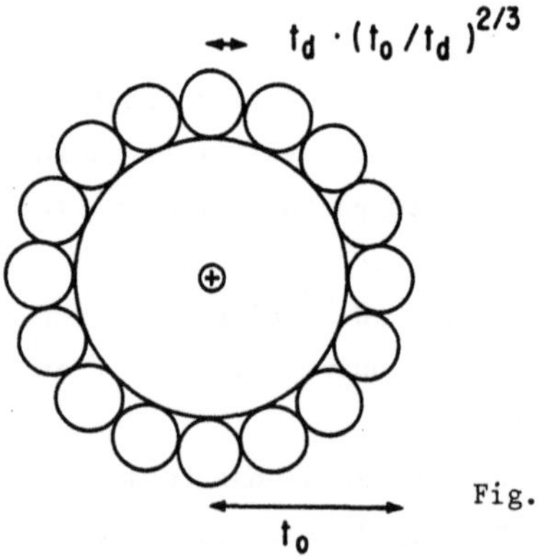

Fig.2 Last scattering

The universe is much flatter at earlier times. Equivalently, the Ω-parameter (7) is fine tuned to unity with great precision. At the Planck epoch of $T \sim 10^{19}$ GeV at which gravity becomes comparable in strength to other elementary particle forces, the degree of fine tuning is astronomical; of order $\sim 10^{-60}$. Unless this sort of fine tuning is realized at the start of the big bang, the universe can not become mature enough to have the present age of $\sim 10^{10}$ y.

Other basic problems of the big bang model are: the smallness of the cosmological constant, $|\Lambda|^{1/2} < 10^{-32}$ eV, lack of understanding of the formation of structures such as galaxies that are believed to have a cosmological origin, and finally, existence of the curvature singularity ($R_{\alpha\beta\gamma\delta}R^{\alpha\beta\gamma\delta} \to \infty$ as $t \to +0$).

II. GUT and Early Universe

The classical big bang model presented thus far is based on microscopic physics up to a distance scale of the nuclear size, of $\sim 10^{-11}$ cm. Recent developments of high energy and elementary particle physics have clarified the microworld to much smaller distances. One of the most crucial questions, how technically possible it is to probe the very early universe, rests on whether the theory enters into a phase of strongly interacting constituents or into a weakly coupled phase. For a strongly interacting system it is virtually impossible, for technical reasons, to attack the early universe. Fortunately for cosmology and remarkably for particle physics, the strong interaction of elementary particle forces becomes weaker and weaker as the distance scale becomes smaller, or equivalently as the energy scale becomes higher. This remarkable property is known as asymptotic freedom, and is a consequence of unbroken SU(3) color gauge theory; quantum chromodynamics (QCD).

While the strong interaction becomes weaker, the electromagnetic interaction grows stronger as the energy increases. The rate of change of coupling "constants" is determined by the renormalization group analysis, and as a result, the strength logarithmically changes. The QCD coupling decreases more rapidly than the fine structure "constant" increases. Eventually the two couplings merge at an energy scale of $\sim 10^{15}$ GeV [6], the unified coupling being of $\sim 1/45$. (Fig. 3)

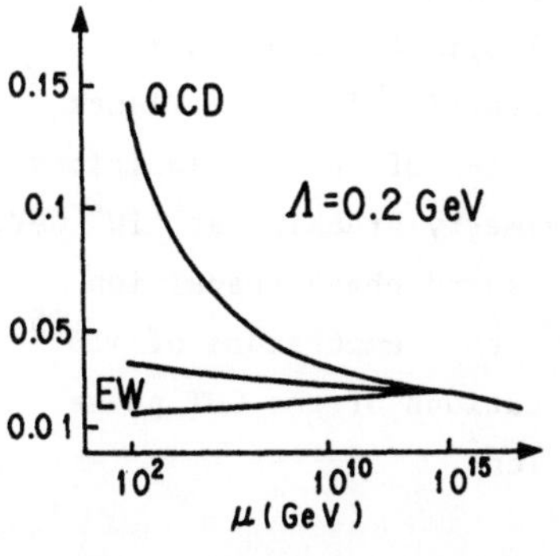

Fig.3 Unified coupling

Unification of coupling constants suggests unification of all elementary particle forces. In this unification picture the apparent disparity of three forces (strong, electromagnetic and weak) is regarded as a low energy manifestation. Electromagnetic and weak interactions are already partially unified in the Weinberg-Salam theory. This partial unification occurs at a scale of $\sim 10^2$ GeV. The grand unified scale is much higher, of $\sim 10^{15}$ GeV. Above this GUT energy scale one can not distinguish three forces: quarks and leptons differ in presence or absence of color, only at low energies. At high energies they behave similarly. Construction of grand unified theories (GUT) thus proceeds as follows: one first chooses an appropriate, simple (in the mathematical sense) group (which must have a rank larger than 4) and assigns a generation of quarks and leptons to an irreducible (or reducible in some cases) representation of the group. The gauge boson belongs to the adjoint representation, but a system of Higgs boson representation, which gives rise to the spontaneous breaking of gauge symmetry, must be specified to yield a heirarchical pattern of symmetry breaking at 10^{15} GeV and 10^2 GeV. Arbitrariness of the choice of Higgs system must be regarded as an ugly feature of GUT at the present level of understanding.

The most successful result of the GUT idea is prediction of the mixing parameter of neutral current; $\sin^2\theta_w \simeq 0.21$, which is in good agreement with low energy data. A remarkable new feature of GUT is baryon nonconservation by the process; $qq \leftrightarrow \bar{X} \leftrightarrow \bar{q}\bar{\ell}$. The X-boson is a new gauge boson in GUT, which couples both to two (anti) quarks ($\overline{qq}$) and to a quark-lepton system ($q\ell$). Since quarks carry 1/3 of the baryon number, this process, $qq \leftrightarrow \bar{q}\bar{\ell}$, changes the baryon number by $\Delta B = \pm 1$. Baryon nonconservation induces nucleon instability by decay processes such as $p \rightarrow \pi^\circ + e^+$, with an estimated lifetime larger than 10^{29} years [7]. Lack of a clear signature of nucleon decay in ongoing experiments indicates [8] that a lower estimate of the lifetime up to $\sim 10^{32}$ years is excluded. The theory is flexible enough to accommodate longer lifetimes, however.

I shall now discuss the most important implication of GUT in the very early universe. GUT and the electroweak unification are based on a spontaneous symmetry breaking of local gauge symmetries; the Higgs mechanism. In general, a spontaneous symmetry breaking is restored at high temperatures, as is well known in condensed matter physics. It can be shown that GUT symmetries are also restored at early epochs of cosmological evolution. Starting from the Planck epoch ($T \sim 10^{19}$ GeV and $t \sim 10^{-44}$ sec), one finds a series of phase transitions; GUT symmetry breaking at $\sim 10^{15}$ GeV and electroweak symmetry breaking at $\sim 10^2$ GeV. The third symmetry breaking has to do with the quark-hadron phase transition related to the quark confinement. For instance, above the temperature of $\sim 10^{15}$ GeV X bosons are as light as W, Z and photons. Implications of the GUT phase transition will be explored in the following two sections.

III. Generation of Baryon Asymmetry

Let us assume that the universe was initially symmetric between baryons and antibaryons, and consider how the asymmetry was created. The initial symmetry can be prepared by the inflation as will be discussed shortly, or by a wash-out process of baryon nonconservation different from the asymmetry-generating mechanism discussed here.

It is not difficult to enumerate necessary conditions for baryon generation [9]. (1) Baryon nonconservation. (2) Violation of particle-antiparticle symmetry, more specifically, violation of C and CP. (3) Departure from chemical equilibrium concerning baryon nonconservation. The first two conditions are obeyed by any decent GUT models, while the last condition of macroscopic nature is given by the expansion of the universe.

The simplest scenario of baryon generation uses the X boson decay [10] of

$$X \to \bar{q}\bar{q} \quad \text{and} \quad q\ell \,, \tag{14a}$$

$$\bar{X} \to qq \quad \text{and} \quad \bar{q}\bar{\ell} \,. \tag{14b}$$

Here q ($\bar{q}$) denotes a quark (antiquark), and ℓ ($\bar{\ell}$) a lepton (antilepton). When a pair of X and $\bar{X}$ decay independently, the baryon number is changed by an amount ε given by

$$\varepsilon = (-\tfrac{2}{3}\gamma_q + \tfrac{1}{3}\gamma_\ell + \tfrac{2}{3}\bar{\gamma}_q - \tfrac{1}{3}\bar{\gamma}_\ell)/(\gamma_q + \gamma_\ell)$$

$$= (\bar{\gamma}_q - \gamma_q)/(\gamma_q + \gamma_\ell) \,. \tag{15}$$

γ_i ($\bar{\gamma}_i$) denotes a partial decay rate of X ($\bar{X}$) into i type modes. By the CPT theorem which must be obeyed by any local quantum field theory, the total width of X and $\bar{X}$ is the same: $\gamma_q + \gamma_\ell = \bar{\gamma}_q + \bar{\gamma}_\ell$. On the other hand, partial rates can differ, $\gamma_q \neq \bar{\gamma}_q$, if C and CP are both violated.

The macroscopic irreversibility operates only if the inverse processes to (14) such as $\bar{q}\bar{q} \to X$ are kinematically blocked. This occurs if the temperature is low enough, $T < m_x$, when the decay starts. The decay time is estimated by equating the decay rate with the expansion rate. Thus, one finds a necessary condition for departure from the chemical equilibrium,

$$\alpha_x m_x \sim \sqrt{N_F}\, T^2/m_{p\ell} \quad \text{with} \quad T < m_x \,,$$

$$\text{namely,} \quad m_x > (\alpha_x/\sqrt{N_F})\, m_{p\ell} \,. \tag{16}$$

A more quantitative estimate of this lower bound has been given in the literature [10]. For the gauge boson of the SU(5) model [11], it is $\sim 10^{15}$ GeV. It is perhaps instructive to think of what might happen if this lower bound is violated. In this case decay and inverse decay will balance until the inverse decay decouples from expansion at a much lower temperature, $T \ll m_x$. By this time abundance of X and $\bar{X}$ have become much suppressed by the Boltzmann factor, making the subsequent baryon generation by the decay process likely to yield too small an asymmetry.

The elementary asymmetry (15) is related to the present baryon-to-photon ratio by a formula like

$$(n_B/n_\gamma)_o \sim 10^{-2} \cdot \varepsilon \ . \tag{17}$$

The dilution factor of 10^{-2} is explained as follows. After baryogenesis is completed, no baryon asymmetry is added, but photons may be created whenever a heavy particle decays or a pair annihilation of heavy particles occurs. What is conserved since the baryogenesis is the baryon-to-entropy density ratio, instead of the baryon-to-photon number ratio. One can then estimate the baryon-to-photon ratio at any temperature by computing the ratio of entropy to photon number at respective temperature. In the present universe this ratio is of $\sim 10^{-2}$ in the SU(5) model.

Perturbative calculation of the asymmetry (15) proceeds as follows. One decomposes the decay amplitude into Born (F_B) and loop (F_L) terms. Difference between X and $\bar{X}$ decays appears in coupling constants, dictated by the CPT theorem, hence

$$\bar{\gamma}_q - \gamma_q = \Sigma\ |g_B^* F_B + g_L^* F_L|^2 - \Sigma\ |g_B F_B + g_L F_L|^2$$

$$= 4\Sigma\ \mathrm{Im}(g_B g_L^*)\ \mathrm{Im}(F_B F_L^*)\ . \tag{18}$$

The summation symbol Σ is meant to sum over final state configurations. CP violation assures in general a nonvanishing complex factor; $\mathrm{Im}(g_B g_L^*) \neq 0$, but another complex factor, $\mathrm{Im}(F_B F_L^*) \neq 0$, is required. A relative phase between a loop and a Born amplitude can be interpreted as a generalized rescattering phase. This phase arises as a result of dynamical interaction among final decay products. What is important to microscopically generate a nonvanishing asymmetry is coexistence of two kinds of phases; one phase that changes its sign when one goes from a particle (X) to its antiparticle ($\bar{X}$), and another phase that does not change.

The total rate in the denominator of (15) is of order $|g_B F_B|^2$, thus the ratio (15) is roughly of order $|g_L/g_B|$. This implies that the magnitude of the baryon asymmetry is suppressed by a higher order coupling; a nice feature to derive the small number of $\sim 10^{-10}$. The calculated asymmetry depends on details of the GUT model. It is not difficult to obtain the required asymmetry, but in simple models it gives interesting constraints [12]. For instance, the simple SU(5) model that was first introduced historically yields too small a baryon-to-photon ratio, while, in the minimal SO(10) model that was later invented, but is esthetically more appealing, interesting correlation with the neutrino mass is found. At present it is best to use the baryon asymmetry as a tool in GUT model building.

IV. Inflation

Nature of the GUT phase transition can have a dramatic effect. The phase transition is characterized by an effective potential V (= free energy at finite temperatures) as a function of the order parameter, the Higgs expectation value ϕ in the case of GUT phase transition. At high temperatures much above the critical value T_c ($\sim m_x$), the potential has a minimum at $\phi = 0$, while at low temperatures much below T_c it has a usual minimum at ϕ_o ($\sim m_x$). If the phase transition proceeds smoothly between these two potentials, it does not much affect the course of cosmological evolution.

On the other hand, in the case of the first order phase transition a local minimum at $\phi = 0$, separated from the global minimum, appears as depicted in the dashed curve of Fig. 4. The transition from the local to the global minimum is made possible either by a thermal fluctuation or by a barrier penetration. But if the height of the barrier remains large until late, the universe is trapped in the metastable phase. During this stage the temperature decreases, obeying the law, $T\, a$ = const. The thermal energy density ρ_{th} thus decreases as T^4, whereas the "vacuum" energy $V(\phi=0)$ remains constant. Thus if the supercooling lasts long enough, the vacuum term overtakes the thermal energy. This causes a major change in the Einstein equation (3). Ignoring the temperature term, one may write evolution equations for the scale factor a and the Higgs value ϕ as

$$\left(\frac{\dot{a}}{a}\right)^2 = \frac{8\pi}{3} G\, V(\phi) \quad , \tag{19}$$

$$\ddot{\phi} + 3\,\frac{\dot{a}}{a}\,\dot{\phi} = -\,\frac{\partial V}{\partial \phi} \quad . \tag{20}$$

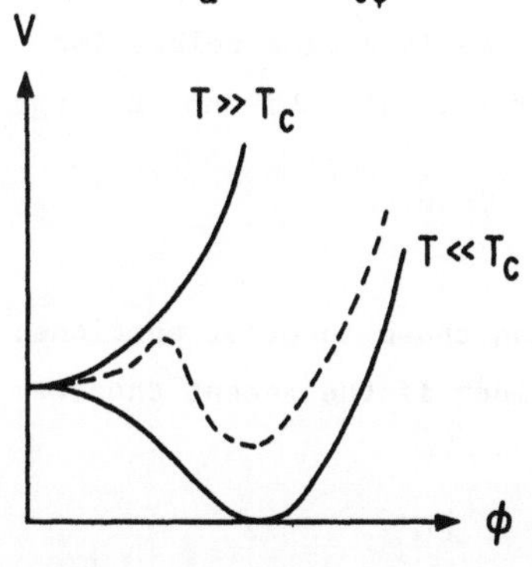

Fig.4 Effective potential

If the potential has a flat part near $\phi = 0$, one may temporarily approximate $V(\phi)$ as a constant $V(0)$, and obtain

$$a(t) \propto e^{Ht} , \quad H = \left(8\pi G\, V(0)/3\right)^{1/2} . \tag{21}$$

Thus the scale factor can exponentially inflate [13].

During this inflation the horizon distance also enlarges,

$$r_H = a(t) \int_0^t dt'\, a^{-1}(t') \sim \frac{1}{H} e^{Ht} . \tag{22}$$

It can be shown that if the e-folding factor $Ht > 65$, the horizon problem is solved. The flatness problem is also solved because

$$\Omega - 1 = k\, \dot{a}^{-2} \propto e^{-2Ht} . \tag{23}$$

In the inflationary picture a small region of causal contact prior to the phase transition can exceed the observable part of the present universe after termination of the GUT phase transition. Much of the information prior to the inflation is diluted essentially to none; in particular the baryon number of the observable part can be extremely small.

Success of the inflation must however explain a smooth exit to a hot Friedmann universe. This is expected to occur toward the end of inflation. The Higgs field ϕ eventually relaxes to the bottom value ϕ_o in the valley of the effective potential. At this bottom the field oscillates with a time scale much smaller than the Hubble scale H^{-1} given by (21). The oscillation of the background field induces a quantum mechanical particle creation via terms of particle fields coupling to the Higgs field. Interaction among created particles usually occurs faster than H^{-1}. This implies that the vacuum energy $V(0)$ is quickly converted into thermal energy among created particles [14]. The baryon asymmetry must be created after this whole process of reheating.

I shall not go into many recent topics [15] related to inflation, but suffice it to say that the field giving rise to inflation need not be the order parameter of the GUT phase transition. What is crucial to inflation is a slow relaxation of a field toward an equilibrium value at zero temperature. There is no shortage of candidate fields in recent unified theories.

V. Higher Dimensional Universe

Topics in this section are far more speculative than those in other sections. I include them to illustrate what one may have to consider if the recent theoretical idea is indeed correct.

A remarkable discovery [16] in recent theoretical developments is the superstring theory that is free of gravitational anomalies and is presumably finite without ultraviolet divergences. This theory makes sense only in a particular spacetime dimension of ten and for a particular set of gauge groups such as $E_8 \times E_8$. An obvious cosmological question one faces is how extra six dimensions compactify to escape detection.

A possible cosmological scenario [17] starts from two equal scale factors, $a_3(t_o) = a_6(t_o)$, associated with a product space of three-and six-dimensional manifolds, $M_3 \times M_6$. For illustration take the three-dimensional manifold M_3 to be a maximally symmetric space of signature k (see (2) for the metric) and M_6 to be 6 - sphere. Suppose that initially the massless ideal gas characterized by the equation of state $p = \rho/9$ (9 = number of space dimensions) fills the entire nine-dimensional space, and both of $a_3(t)$ and $a_6(t)$ grow like $t^{1/5}$ (generalization of RD Friedmann solution) as time goes on. Two scales, however, get separated by difference in curvature terms

$$\ddot{a}_i/a_i = - k_i(i-1)\, a_i^{-2} + \cdots .$$

One can readily prove [17] that a space with a larger dimensionality expands less, namely, $a_3(t) > a_6(t)$. Once separation occurs, it is accelerated by gravity, and results in what is called Kasner solution; $a_i(t) \propto (t_1 - t)^{\gamma_i}$ with $3\,\gamma_3 + 6\,\gamma_6 = 3\,\gamma_3^{\,2} + 6\,\gamma_6^{\,2} = 1$. This solution is characteristic of the vacuum metric in which curvature and matter terms are negligible. The exponents satisfy an inequality, $\gamma_3 < 0 < \gamma_6$, hence the six dimensions shrink with a power law. At some stage of the Kasner collapse the temperature becomes much less than a_6^{-1}; deexcitation of 6 - space. The collapse is expected to be halted (Fig. 5) by some kind of pressure forces such as the gravitational Casimir pressure [18] or a gauge field condensate [19].

Cosmological compactification remains largely a subject of speculation. It is perhaps more important to contemplate how one can detect the presence of the extra dimensions. Right after the collapse the extra size $a_6(t)$ undergoes a damped oscillation, which then induces particle creation. The baryon asymmetry may have been generated by subsequent X decay [20]. The damped oscillation also causes time variation [20], [21] of the effective four-dimensional coupling

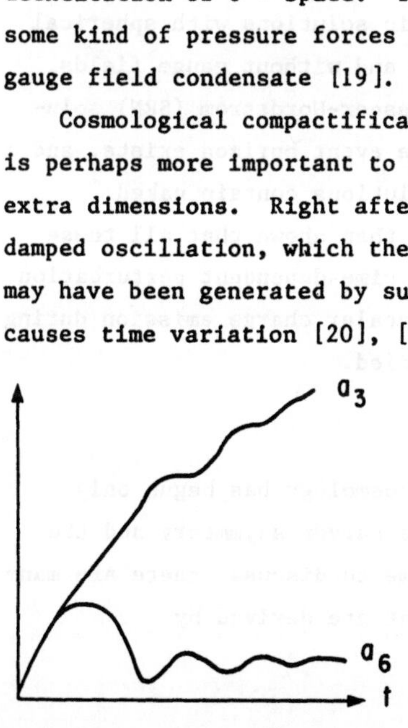

Fig.5 Compactification

"constants" such as the gravitational and the fine structure constant. Unless the extra space compactifies to a Ricci flat space, time variation, however, appears too small to have observational possibilities.

Much more dramatic effects can occur if the Kaluza-Klein idea of compactification remains valid, namely if the compact space admits an isometry such that a part of observed gauge symmetries arises as a result. (A symmetry of the compact extra space induces a local gauge symmetry in the four-dimensional sector in the old Kaluza-Klein theory. This may or may not be the case in superstring theories.) Prior to the compactification the universe may then pass through a stage in which the extra space is deformed, violating some of those gauge symmetries. The electromagnetic gauge invariance, hence the electric charge conservation, can be violated during this stage, to yield an excess electric charge [22]. There is however no reliable way to estimate this charge excess. Moreover, various cosmological considerations severely restrict the amount of leftover charge [22].

In principle the presence of the extra space may have important effects on the black hole formation. It is well known by the no-hair theorem that the black hole with event horizon can only have three kinds of labels; mass, angular momentum and electric charge. On the other hand, the logarithm of the extra size factor, $\ln a_6$, acts as a scalar field, viewed as a four-dimensional field theory. Variation of the size factor implies existence of a scalar charge and is not allowed beyond the event horizon. Since the extra size can vary before gravitational collapse, one is led to ask how a scalar charge emission occurs during the collapse. I recently made two contributions to this problem. First, in the special case of Ricci flat compactification all static solutions with spherical symmetry have explicitly been constructed [23], with and without gauge fields. Except in two special cases of the Schwarzschild-Reissner-Nordstrom (SRN) solution in which the size is everywhere constant and the event horizon exists, and a completely regular wormhole solution, all generic solutions contain naked singularity not covered by the event horizon. It is then shown that all these singular solutions, except SRN, are unstable against time-dependent perturbation. The perturbative instability presumably implies the scalar charge emission during the black hole formation, but much remains to be studied.

VI. Conclusion

A cosy interaction between particle physics and cosmology has begun only recently. We now have a new way of understanding the baryon asymmetry and the cosmological principle. Although we did not have time to discuss, there are many interesting and useful bounds on particle physics that are derived by cosmological/astrophysical arguments [15].

A major cosmological problem of immediate concern [15] is nature of the dark matter that prevails on various mass scales. At a more fundamental level the following problems need to be elucidated: initial singularity, cosmological constant, and quantum gravity. It seems that a better understanding of these deep problems must await a major breakthrough of more fundamental microscopic theories such as the superstring theory.

References

1. For a review, A.M. Boesgaard and G. Steigman, Ann. Rev. Astr. Astrophys. 23 (1985) 319.
2. J. Yang, M.S. Turner, G. Steigman, D.N. Schramm and K.A. Olive, Ap. J. 281 (1984) 493.
3. For a review, G. Steigman, Ann. Rev. Astr. Astrophys. 14 (1976) 339.
4. P.M. Lubin, G.L.Epstein and G.F. Smoot, Phys. Rev. Lett. 50 (1983) 616; D.J. Fixsen, E.S. Cheng and D.T. Wilkinson, Phys. Rev. Lett. 50 (1983) 620.
5. J.M. Uson and D.T. Wilkinson, Nature 312 (1984) 427.
6. H. Georgi, H.R. Quinn and S. Weinberg, Phys. Rev. Lett. 33 (1974) 451.
7. J. Ellis et al., Nucl. Phys. B176 (1981) 61, and references therein.
8. Y. Totsuka, in Proceedings of the 1985 International Symposium on Lepton and Photon Interactions at High Energies, ed. by M. Konuma and K. Takahashi (1986).
9. A.D. Sakharov, Sov. Phys. J.E.T.P. Lett. 5 (1967) 24; M. Yoshimura, Phys. Rev. Lett. 41 (1978) 281, 42 (1979) 740(E); A.Y. Ignatiev, N.V. Krasnikov, V.A. Kuzmin and A.N. Tavkhelidze, Phys. Lett 76B (1979) 594. For a review, M. Yoshimura, in "Grand Unified Theories and Related Topics" (World Scientific, Singapore 1981).
10. D. Toussaint, S.B. Treiman, F. Wilczek and A. Zee, Phys. Rev. D19 (1979) 1036; S. Weinberg, Phys. Rev. Lett. 42 (1979) 850; M. Yoshimura, Phys. Lett. 88B (1979) 294.
11. H. Georgi and S.L. Glashow, Phys. Rev. Lett. 32 (1974) 438.
12. Review by M. Yoshimura, Ref 9 .
13. A.H. Guth, Phys. Rev. D23 (1981) 347; A.D. Linde, Phys. Lett. 108B (1982) 389; A. Albrecht and P.J. Steinhardt, Phys. Rev. Lett. 48 (1982) 1220.
14. A. Albrecht, P.J. Steinhardt, M.S. Turner and F. Wilczek, Phys. Rev. Lett. 48 (1982) 1437.
15. For a review, M. Yoshimura, in Proceedings of 23rd International Conference on High Energy Physics, Berkeley 16-23 July, 1986.
16. For an introductory review, K. Kikkawa, in this Proceedings.
17. For a review, M. Yoshimura, in Proceedings of Takayama Workshop on Toward Unification and its Verification, KEK-TH-107 (1984).
18. T. Appelquist and A. Chodos, Phys. Rev. Lett. 50 (1983) 141; P. Candelas and S. Weinberg, Nucl. Phys. B237 (1984) 397; T. Koikawa and M. Yoshimura, Phys. Lett. 155B (1985) 137.
19. P.G.O. Freund and M.A. Rubin, Phys. Lett. 97B (1980) 233; S. Randjbar-Daemi, A. Salam and J. Strathdee, Nucl. Phys. B214 (1983) 491.
20. T. Koikawa and M. Yoshimura, Phys. Lett. 150B (1985) 107 and KEK-TH91 (Revised).
21. Y.S. Wu and Z. Wang, Phys. Rev. Lett. 57 (1986) 1978.
22. S. Orito and M. Yoshimura, Phys. Rev. Lett. 54 (1985) 2457.
23. A. Chodos and S. Detweiler, Gen. Rel. Grav. 14 (1982) 879; M. Yoshimura, Phys. Rev. D34 (1986) 1021; T. Koikawa and M. Yoshimura, Phys. Lett. 189B (1987) 29.

Gravitational Collapse

T. Nakamura
Department of Physics, Kyoto University, Kyoto 606, Japan

1. Gravitational Instability

There are many kinds of structures in the universe. There are, for example, neutron stars, white dwarfs, main sequence stars like the sun, planets, molecular clouds, globular clusters, elliptical galaxies, spiral galaxies, clusters of galaxies, super clusters of galaxies and so on. The size of these structures ranges from $10^5 cm$(neutron stars) to $10^{26} cm$ (super clusters) while the density varies from $10^{15} g/cm^3$ to $10^{-29} g/cm^3$. There is a difference of 44 orders of magnitude in density between neutron stars and superclusters of galaxies. Although there are four kinds of fundamental interactions in our universe which are responsible for the formation of these structures, gravity is the most important because the range of the force is long compared with the strong and weak interactions and because there is no shielding distance in contrast to the electromagnetic interaction. These two characteristic of gravity are by far the most important for the formation of the various structures in the universe.

As an example let us consider the density perturbation in a uniform, fully ionized plasma in which we consider only the electrostatic interaction and that in a uniform neutral hydrogen gas in which we consider only Newtonian gravitational interaction to clarify the difference between electromagnetic and gravitational interactions. In plasma, we have the plasma oscillation and its dispersion relation is written as

$$\omega^2 = k^2 c_s^2 + \frac{4\pi n_0 e^2}{m_e},$$

where c_s and n_0 are the sound velocity and the density of the plasma, respectively. The dispersion relation clearly shows that ω^2 is real for all wave numbers k. This means that the small density perturbation in the fully ionized plasma does not grow exponentially and the system is stable. This is essentially due to the repulsive force acting between electrons. On the other hand, if we consider the oscillation in a self-gravitating neutral gas cloud which may exist in interstellar space, the corresponding dispersion relation becomes

$$\omega^2 = k^2 c_s^2 - 4\pi G \rho_0,$$

where ρ_0 and G are the gas density and the gravitational constant, respectively. This dispersion relation is formally obtained by replacing $n_0 e^2/m_e$ by $-G\rho_0$ in the previous relation for plasma.

This is due to the fact that both electric and gravitational interactions obey the inverse square law. The change of the sign comes from the fact that the gravitational interaction is not repulsive but attractive. This difference yields an essential difference of the nature of the small density perturbation. Contrary to the plasma case, for $k \leq k_J \equiv \sqrt{4\pi G\rho_0}/c_s$ ω^2 becomes negative. This means that the small density perturbation grows exponentially and that the self-gravitating uniform density gas cloud is unstable. If we consider the mass (M) of the gas within a sphere with a radius $2\pi/k$, we can re-express the condition of the above instability as the perturbation with mass greater than the critical mass $M_{cr} \equiv 4\pi/3 \cdot \rho_0 \cdot 2\pi/k_J{}^3$. This critical mass is called the Jeans mass and the instability itself is called the Jeans instability. The Jeans instability is the most important phenomena relevant to the formation of the various structures in the universe and is usually called Jeans instability. If there is a density perturbation satisfying $k \leq k_J$, the perturbation grows and gravitational collapse of the system ensues.

2. Neutron Stars and Black Holes

Almost all the astronomical objects are rotating. The earth is rotating once per day and the speed of rotation on the equator is 460m/s. The sun revolves in a period of about 25 days and the rotation speed is 2km/s. The earth rotates not only on its axis but also around the sun. The speed of the revolution is about 29km/s. Furthermore the solar system with the sun at its center belongs to the Galaxy which consists of a hundred billions of stars and is revolving around its center, which is thirty thousand light years far away from us, with a rotation speed of 250km/s once every quarter of a billion years.

In general, a star not only revolves around the center of our Galaxy but also rotates on its axis. Observationally, the more massive the star, the faster it is rotating. The fastest rotation speed is ~ 400km/s, which is close to the escape velocity(~ 600km/s). As these examples show, almost all astronomical objects are rotating, that is, they have angular momentum. So in the gravitational collapse discussed in Sect. 1, the effect of angular momentum is usually very important.

In 1982, an astronomical object which is rotating faster than any other object in the universe was discovered. The object is called a superfast pulsar or a millisecond pulsar. This pulsar emits radio waves like the search light of a lighthouse with 1.5 millisecond period. That is, the radio wave turns on and off 400 times every second. The pulsar itself is considered to be a spinning neutron star. In a neutron star the density of matter is condensed up to 10^{15}g/cm^3 so that protons capture electrons to become neutrons to reduce the Fermi energy of the electrons. The mass of a typical neutron star is $1.4M_\odot (1M_\odot = 2 \cdot 10^{33}$g) and its radius is about 10km. Thus the rotation speed of the superfast pulsar becomes 30,000km/sec, which is a tenth of the light velocity. At such a high velocity, one might think that no equilibrium state of the star exists . However as the radius of the star is very small, the gravitational force is much stronger than the centrifugal force and the escape velocity becomes three tenths of the light velocity. So there is a gravitational equilibrium even for such a rapidly rotating system.

At present, about 400 radio pulsars are known to exist. The existence of the pulsar was discovered in 1967. However the existence of the neutron star itself was predicted by Oppenheimer and Volkoff in 1939[1]. When gravity is very strong as it is in neutron stars, the effect of general relativity is very important. They studied the spherically symmetric equilibrium solutions of neutron stars by neglecting the strong interactions of neutrons for simplicity. As a result they found that there is no equilibrium solution for a neutron star if the gravitational mass exceeds 0.75 $M_{\odot}$. This is because the pressure of neutrons cannot sustain the strong general relativistic gravity if the mass of the star is large enough. We usually call this maximum stable mass the critical mass of the neutron star. In reality we must take into account the effect of the strong interaction of neutrons, which is not well known for densities greater than the nuclear density $3 \cdot 10^{14}\mathrm{g/cm}^3$, to study the equilibrium structure of neutron stars. However the existence of the critical mass can be proved under general assumptions irrespective of the details of the equation of state above the nuclear density. If one assumes that:

1) Einstein's general theory of relativity is correct.

2) The sound velocity in matter is smaller than the light velocity.

3) The star is spherically symmetric ,

then one can show that the critical mass is at most 3 $M_{\odot}$. The critical mass of neutron stars has been calculated by many authors using various equations of state. It ranges from $1.4 M_{\odot}$ to $2 M_{\odot}$. Even if one includes the rotation of the star, the critical mass increases at most by a factor of 1.5. So the critical mass of the neutron star exists and its value should be smaller than 4.5 $M_{\odot}$

At this stage it is natural to ask: What is the fate of a neutron star if it has a mass greater than the critical mass? As the pressure cannot sustain the gravity in this case the star should contract without a limit. Oppenheimer and Snyder pointed out this catastrophic collapse of the star in their paper entitled *On Continued Gravitational Contraction* in 1939[2]. This collapsed object is now called a *black hole.* If the system is spherically symmetric, a black hole is described by a Schwarzschild solution (1916[3]) which is a spherically symmetric vacuum solution of the Einstein equation. (Here we mean by *vacuum* that the density of the matter is zero but the gravitational fields exist.) So we usually call this black hole a Schwarzschild black hole. The only parameter that characterizes the Schwarzschild black hole is the mass. In 1917, Weyl discovered a series of axially symmetric vacuum solutions of the Einstein equations. In Weyl's solution there is no angular momentum in the system but the gravitational field at large distance has multipole components other than a monopole term. These multipoles are created by cylinder-like or disk-like singularities of the space- time near the origin.

In 1963[4], Kerr found a new axially symmetric vacuum solution of the Einstein equation. The important characteristic of the Kerr solution is that the system has angular momentum. The Kerr solution is characterized by two parameters, namely, the mass and the angular momentum. When the angular momentum is zero, the Kerr black hole becomes the Schwarzschild black hole. In 1972[5], Tomimatsu and Sato discovered a new series of solutions called the TS solution. The TS solution is characterized by three parameters, the mass, the angular momentum and the deformation. After the discovery of the TS solution, various mathematical methods to obtain the axially symmetric vacuum solution of the Einstein equation were developed. As a result it is now possible to obtain the solution algebraically with any number of parameters.

In the above paragraph the difference between the black hole and the axially symmetric vacuum solution of the Einstein equation has not been clarified. One of the important features which distinguishes them is the event horizon. When the star collapses general-relativistically, an extremely strong gravitational field is formed. It is then possible that the gravitational field becomes so strong that even an outgoing light ray cannot escape from the star. The event horizon is defined by the boundary of the region from which even light rays cannot escape. So we have no causal relation to the region inside the event horizon. Another important idea is the singularity in space-time. Intuitively, the singularity is defined by the space-time points in which the tidal force or the density diverges. This means that at the singularity the Einstein equation is violated. We must use a quantum theory of gravity. However, such a theory has not yet been established.

All the known axially symmetric vacuum solutions of the Einstein equation have singularities. But the Schwarzschild and the Kerr solutions are unique in the sense that the singularity is covered by the event horizon. This kind of structure is very convenient because we cannot look at the singularity. The existence of the singularity means the violation of the basic equation, and the physics there is governed by a theory which we do not know. However, if the singularity is covered by the event horizon, we have nothing to do with what happens near the singularity. If we take a position that we treat in physics only phenomena which we can see in principle, we can consider the Einstein theory of relativity as the true theory so long as we consider the space-time outside the event horizon. Thus there is no contradiction between the existence of the space-time-like Kerr solution and the Einstein theory.

One might think that there would be many kinds of solutions which have a black hole structure, that is, the singularity covered by the event horizon. However surprisingly, Israel, Carter, Hawking and Robinson[6] proved that the Kerr solution is the unique axially symmetric vacuum solution of the Einstein equation in which all the singularities are covered by the event horizon. The assumption that all the singularities should be covered by the event horizon is usually called the cosmic censorship hypothesis (Penrose 1973[7]). So now in the standard view point of the gravitational collapse, a black hole means a Kerr black hole. (Note that the Schwarzschild black hole is a special case of the Kerr black hole.) This result is in a sense very strange. As we discussed in the previous section most astronomical objects are rotating. A black hole is also expected to have an angular momentum, which in itself is not strange. But what is strange is the fact that the Kerr black hole is characterized by only two parameters, whereas a rotating normal star must be characterized by more than two parameters.

3. Numerical Relativity

Figure 1 shows the standard view point of the gravitational collapse of a rotating star to a Kerr black hole by Penrose(1973[7]). The energy source of the star is the nuclear energy. In the first stage of the evolution of the star a proton-proton chain reaction proceeds and heliums are formed from the fusion of four protons. The star acquires the energy from this reaction and emits optical photon from the photosphere. As the pp reaction proceeds the central part of the star becomes a helium-rich core. Then in the next phase a carbon is formed from three heliums, that is, the helium burning begins, and this is followed by the carbon burning and so on. Finally the central core of the star becomes iron.

As the binding energy of iron is the largest, no nuclear energy is available at this stage. Then it becomes impossible to sustain the star against gravity and the gravitational collapse ensues. If the mass of the collapsing core is larger than the critical mass of the neutron star, the gravitational collapse is not halted and finally there appears a trapped region in which gravity

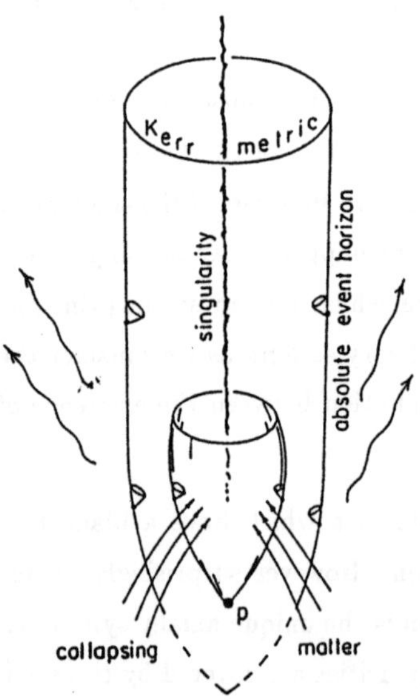

FIGURE 1. Standard viewpoint of gravitational collapse of a rotating star to a black hole.

is so strong that even light rays cannot escape. In Figure 1, the wavefront of the outgoing light rays from a point P at first expands but finally it contracts due to the strong gravity. The outer boundary of such a region is called an event horizon. As gravity increases, the acceleration of the matter also increases. Then the distortion of the space-time (or tidal force) begins to propagate as a wave, just as the electromagnetic wave is emitted in association with the acceleration of the charged particles. We call this wave a gravitational wave. After the distortion of the space-time has been radiated as gravitational waves, the final stationary space-time is left. In the standard view point of the gravitational collapse this final structure of the space-time should be a Kerr black hole.

Numerical relativity is a branch of the study of the general theory of relativity in which one tries to clarify the physics in strong gravity by solving the Einstein equation numerically as an initial value problem for processes like Figure 1. In a narrow sense numerical relativity was born in 1976[10]. However the first work on the collapse of matter to a black hole was performed by Oppenheimer and Snyder (1939[2]) as mentioned in the previous section. They could solve the problem analytically because they used pressureless matter which is unrealistic. If one assumes a realistic equation of state it is impossible to solve the problem analytically even for spherically symmetric space-times. As one needs a computer to solve the problem of gravitational collapse, the progress of numerical relativity depends on the speed and the size of memory of the computer. Computers in the 1960s enabled us to simulate the general relativistic collapse of the spherically symmetric stars. Assuming that all the quantities depend only on r and t, May and White simulated the collapse of stars of mass $1.1 M_{\odot}$ and $2 M_{\odot}$ in 1966[8]. Matsuda and Sato solved processes of formation of supermassive ($\sim 10^8 M_{\odot}$) black holes in 1969[9]. In spherically symmetric space-time the vacuum solution of the Einstein equation should be a Schwarzschild solution by Birkoff's theorem. So they obtained the Schwarzschild black holes finally.

Computers in the 1970s enabled us to simulate the general relativistic collapse of axially symmetric stars to black holes. Axial symmetry means that if we adopt the cylindrical coordinates (R,φ,Z) all the quantities are independent of φ. In 1976 Smarr[10] at Illinois university simulated the collision of two black holes. At first two black holes are located on the symmetry axis at rest. They begin to attract each other by the gravitational force and finally they coalesce to form a single Schwarzschild black hole. His calculation is epoch-making in the sense that the simulation was the first numerically generated non-spherical dynamical space-time with an asymptotic flatness.

4. General Relativistic Collapse of Axially Symmetric Rotating Stars

What do we want to know by calculating the collapse of rotating stars? Two of the most important questions are:

1) What is the final structure of the space-time after the collapse of rotating stars?

2) What kind of information can we extract from the gravitational radiation emitted during the collapse? By observing gravitational waves in future what kind of physics in strong gravity can we know?

Before the collapse there can be various initial conditions. For example there are many kinds of rotation laws: a star may rotate rigidly or the angular velocity at the center may be much faster than that at the surface. What is the density distribution? It may be either homogeneous or centrally condensed. Concerning the distributions of angular momentum and mass we have in principle two arbitrary functions as initial conditions. Equations of state are also different for different mass of the stars. So we have three arbitrary functions to specify each collapse. If the final results strongly depend on the details of these initial conditions, we will only have a large amount of computer data and it will be difficult to discuss the physics. However if they do not, we will be able to predict the fate of the star qualitatively once we know the initial conditions.

Let us define a nondimensional quantity q by

$$q = \frac{\text{Total angular momentum of the star}}{M \cdot GM/c^2 \cdot c}$$

where M and c are the mass of the star and the light velocity, respectively. The quantity q is calculable when the initial condition is given and it remains constant afterwards. If the value of q is unity, the rotation speed becomes of the order of the light velocity when the star approaches the black hole because the radius of the black hole is given by $2GM/c^2$(= 3km for $M = M_\odot$). What is the value of q for astronomical objects? For example, the value of q for the sun is 0.18. This does not mean that the sun will become a neutron star or a black hole but since q depends only on the mass and the total angular momentum, it is possible to define the value of q for the sun. The value of q for the superfast pulsar is 0.23 and the massive stars which are responsible for the formation of neutron stars and black holes are believed to have q greater than 1. One might think it strange that the sun and the superfast pulsar have almost the same value of q. This is because while the rotation velocity of the sun is four orders of magnitude smaller than that of the superfast pulsar, the radius of the sun is four orders of magnitude greater than that of the pulsar. Note that the angular momentum is (mass)·(radius)·(velocity).

In the second section, we mentioned that the Kerr black hole is the unique one under the cosmic censorship hypothesis and that it is characterized by only two parameters,the mass (M) and the angular momentum (J). It is also possible to define q for the Kerr black hole since q is calculable once the mass and the angular momentum of the system are known. Can the Kerr black hole have any value of q? The answer is No. The maximum value of q for the Kerr black hole is unity. If q is greater than unity, then the singularities are no longer covered by the event horizon. This is because for $q \geq 1$ the event horizon disappears. This situation may have something to do with our intuition that for q greater than unity the rotation velocity of a star with radius GM/c^2 exceeds the light velocity.

Now let us consider the general relativistic collapse of a star of mass greater than the critical mass of neutron stars. If the star is not rotating, the final fate is definitely a Schwarzschild black hole. However if it is rotating very fast with a value of q greater than unity, which is plausible for massive stars, what then is the final fate of the star? If it becomes a black hole its angular momentum is too large for it to be a Kerr black hole. Then the cosmic censorship hypothesis should be violated and we will see the formation of singularities which are not covered by the event horizon. If the cosmic censorship is correct, then what happens for this star with q greater than unity? Under the cosmic censorship hypothesis, if q is smaller than 1, the Kerr black hole should be formed irrespective of initial conditions. Is this really the case?

To answer the above questions many heavy numerical simulations of the general relativistic collapse of rotating stars have been performed by Nakamura and Sato (1981-1984[11]) as well as Stark and Piran(1985[12]). These calculations are unique in the sense that the problem they treated is more realistic than the two black hole collision. Numerical simulations show that there is a critical value of q $(= q_c)$ for each series of calculations. If q is smaller than q_c, a black hole is formed (i.e. the formation of an event horizon), while if q is greater than q_c, a black hole is not formed but instead an expanding ring or an expanding disk depending on the initial conditions. The Kyoto group as well as Stark and Piran performed five series of numerical simulations. In each series of calculations the distribution of density and angular momentum as well as the equation of state are different. As a result they found that q_c ranges from 0.8 to 1.2. This result suggests that q_c is almost independent of the equation of state, initial rotation laws, initial density distribution. If we consider numerical relativity as an experiment using a computer, an experimental value of q_c is 1.0 ± 0.2. This suggests that the cosmic censorship hypothesis holds under plausible initial conditions such as density decreasing as a function of the radius, which should be the case for the massive stars.

If q is greater than unity, the final results strongly depend on the initial conditions. For example, in general relativistic collapse of neutron stars with rotation , depending on the rotation laws, oblate , disk-like or ring-like neutron stars are formed instead of black holes for $q \geq 1$. The shapes of these stars suggest that they are unstable with respect to fragmentation because for example a thin disk with an aspect ratio greater than 2π is known to be unstable. As early type stars, which are responsible for the formation of neutron stars and black holes, are rotating fast, the value of q of collapsing cores is expected to be greater than unity (de Felice 1986). Therefore the general relativistic collapse with $q \geq 1$ will not be a rare event but in fact quite common.

Nakamura and Sato (1982) have also investigated the axially symmetric collapse of non-rotating deformed stars. They used the equation of state in which the sound velocity quickly approaches the light velocity above the nuclear density. This means that they used the hardest equation of state. They gave some initial oblate-like or prolate-like deformation to the stars. They found that only when the deformation is large and the initial internal energy is small, are there no event horizons but instead, bar-like or disk-like singularities which resemble the Weyl solution. However, a small internal energy and a large deformation without rotation are unrealistic. So their result suggests that under realistic initial conditions a Schwarzschild black hole is usually formed for the collapse of non-rotating axially symmetric stars. Another interesting series of simulations has been performed by Nakamura (1984) . He considered the collapse of non-rotating supermassive stars with poloidal magnetic fields. He found that unless the initial magnetic field is too large, event horizons are formed.

In conclusion, one of the important results in the general relativistic collapse of axially symmetric stars is that numerical simulations tell us that under the plausible initial conditions in realistic astrophysical situations, Kerr black holes are usually formed instead of naked singularities and the cosmic censorship hypothesis seems to be valid(Nakamura et al. 1987[13]).

5. Determinant of Rotating Collapse

If a system has rotation or magnetic field and it begins to gravitationally collapse, then it is expected that the shape of the star will be disk-like because the contraction along the rotational axis (or along the magnetic fields) is much easier than that perpendicular to the axis. The linear stability of such a disk was studied by Goldreich and Lynden-Bell in 1965[15]. They found that the system is gravitationally unstable if

$$G\rho_0 \geq \Omega_0^2$$

where ρ_0 and Ω_0 are the central density of the disk and the angular frequency of the disk, respectively. They showed also that the above criterion depends on the equation of state very weakly. If the above criterion is satisfied the disk will fragment into many pieces.

In the Newtonian regime, several authors have already performed three-dimensional numerical simulations of isothermal rotating clouds to study such a fragmentation process in relation to the formation of proto stars. According to these one of the important parameters which determines whether the cloud fragments or not is

$$q' = \frac{\text{Total angular momentum of the cloud}}{M \cdot GM/c_s^2 \cdot c_s}$$

where c_s and M are the sound velocity and the mass of the cloud, respectively. In the previous section we showed that q defined in Sect. 4 is the determinant of the general relativistic collapse of axially symmetric rotating stars to black holes. If one compares the definitions of q and q', one sees that they are essentially the same. That is if we replace c by c_s in the definition of q, we obtain q'. Although the Newtonian gravity and the general relativistic gravity are different, because the rotation is a common factor in both problems, the determinants of the systems become the same thing. Then we can expect that if we choose the velocity in the definition of q properly, then q becomes the determinant of the collapse of rotating stars or clouds in general. So if we calculate the value of q at t=0, then we can predict the fate of the system qualitatively.

If q is greater than unity in the general relativistic collapse of rotating stars it is expected that a disk will be formed. However such a disk is unstable against fragmentation according to the criterion by Goldreich and Lynden-Bell if the density of the disk becomes large, i.e., if the thickness of the disk becomes small. For this situation the assumption of axial symmetry is not good. We must treat the general system, that is, all the quantities depend on x,y,z and t. Computers available in the late 1980s will enable us to simulate such a general relativistic fragmentation process and the emission of gravitational waves from the collapse and an important theme in numerical relativity is to find invariant relations or parameters like q for non-axially symmetric cases.

References

1) J.R. Oppenheimer and G. Volkoff : Phys. Rev. 55, 374(1939)
2) J.R. Oppenheimer and H. Snyder : Phys. Rev. 56, 455(1939)
3) K. Schwarzschild : Sitzber. Deut. Akad. Wiss. 189(1916)
4) R.R. Kerr : Phys. Rev. Lett. 11, 237(1963)
5) A. Tomimatsu and H. Sato : Phys. Rev. Lett. 29, 1344(1972)
6) W. Israel : Phys. Rev. 164, 1776(1967) ;
B. Carter : Phys. Rev. Lett. 26, 331(1971);
D.C. Robinson : Phys. Rev. Lett. 34, 905(1975)
7) R. Penrose : I.A.U. Symp. 64, 82(1973)
8) M.M. May and R.H. White : Phys. Rev. 141, 1232(1966)
9) T. Matsuda and H. Sato : Prog. Theor. Phys. 41, 1021(1969)
10) L. Smarr : Ann. N.Y. Acad. Sci. 301, 569(1977)
11) T. Nakamura and H. Sato : Prog. Theor. Phys. 65, 1876(1981); Ann. N.Y. Acad. Sci. 422, 56(1984); Phys. Lett. 86A, 318(1981); Prog. Theor. Phys. 66, 2038(1981); Prog. Theor. Phys. 67, 1396(1982); Prog. Theor. Phys. 71, 1144(1983)
12) R.F. Stark and T. Piran : Phys. Rev. Lett. 55, 891(1985)
13) T. Nakamura et. al. : Prog. Theor. Phys. Supple. 90(1987)
14) P. Goldreich and D. Lynder-Bell, M.N.R. Ast. Soc. 130, 97(1965)

Solar Neutrinos

H. Morinaga

Physics Department, Technical University of Munich,
D-8046 Garching, Fed. Rep. of Germany

Efforts to detect the neutrinos which are emitted from the sun in the course of energy production due to nuclear reactions are discussed. Special attention is given to progress in attempts to detect the very low energy, but very abundant component, so-called pp neutrinos by means of the radiochemical detection of ^{71}Ge and the geochemical detection of ^{205}Pb.

I. Solar Neutrino Flux and its Energy Spectrum

1) Solar Energy Source

We know from various sources, especially from the evolution of biological systems on earth, that the sun has emitted thermal energy for a very long time at a constant rate. The temperature of the earth's surface at the time of the dinosaur was not much different from now. The existence of life began more than 30 million years ago. The sun is apparently in a steady state with the loss of the thermal energy being compensated by some energy source. The gravitional energy due to the gravitational contraction is not enough to account for such energy loss over a long period of time.

2) Thermonuclear Reactions in the Sun

Bethe was the first to mention the possibility of thermonuclear reaction at the center (the hottest spot) of the sun. The energy comes from the reaction of protons, which produces α particles. Specifically, the so-called CNO cycle was proposed. The CNO cycle runs as follows

$$^{12}C + p \rightarrow {}^{13}N + \gamma$$
$$^{13}N \rightarrow {}^{13}C + e^+ + \nu_e$$
$$^{13}C + p \rightarrow {}^{14}N + \gamma$$
$$^{14}N + p \rightarrow {}^{15}O + \gamma$$
$$^{15}O \rightarrow {}^{15}N + e^+ + \nu_e$$
$$^{15}N + p \rightarrow {}^{12}C + \alpha$$

or

$$^{15}N + p \rightarrow {}^{16}O + \gamma$$
$$^{16}O + p \rightarrow {}^{17}F + \gamma$$
$$^{17}F \rightarrow {}^{17}O + e^+ + \nu_e$$
$$^{17}O + p \rightarrow {}^{14}N + \alpha$$

In short, four protons are lost in a cycle, one α-particle is produced along with 28 MeV of heat and two neutrinos.

Nowadays after more detailed examination, the main energy source is considered to be the so-called pp chain which runs as follows:

		%	$E_{e\,max}$
pp	$p+p \rightarrow d+e^{+}+\nu_e$	99.75	0.42
or			
pep(pH)	$p+p+e \rightarrow d+\nu_e$	0.25	1.44*
	$d+p \rightarrow {}^{3}He+\gamma$		
ppI	${}^{3}He+{}^{3}He \rightarrow \alpha+2p$	86	
	or		
	${}^{3}He+\alpha \rightarrow {}^{7}Be+\gamma$	14	
ppII	${}^{7}Be+e^{-} \rightarrow {}^{7}Li+\nu_e$	90	0.86*
	${}^{7}Li^{*}+\nu_e$	10	0.38*
ppIII	${}^{7}Be+p \rightarrow {}^{8}B+\gamma$	0.02	
	${}^{8}B \rightarrow {}^{8}Be+e^{+}+\nu_e$		14.1
	${}^{8}Be \rightarrow 2\alpha$		

where % gives the branching calculated by Bahacall [1] using the standard solar model. (* is the case where the neutrino has a line spectrum.)

3) Neutrino Flux and the Energy Spectrum

The number of neutrinos per cm^2 per sec at the surface of the earth is the subject of an elementary calculation. Since we know the energy flux (2 cal/cm^2 sec) and the number of neutrinos is 1 per 28 MeV/2, it amounts to 6×10^{10} neutrinos/cm^2 sec.

The spectrum of the neutrino is, on the other hand, not that simple. The total spectrum is the sum of all the participating neutrino-producing processes. If a neutrino is produced together with a positron, its spectrum is a continuous Fermi-type spectrum. If a neutrino is produced without a positron, i.e., by annihilating an electron, it has a line spectrum. The amount of each reaction type is strongly dependent on the temperature. The ratio of different processes from the standard solar model is given in the previous section.

4) Deformation of the Initial Neutrino Spectrum Due to the Mikheyev-Smirnov Effect

Quite recently, Mikheyev and Smirnov noted the possibility that the original electron neutrino (ν_e) spectrum may be seriously deformed after the electron neutrinos have travelled through a dense electron atmosphere before finally leaving the surface of the sun [2]. Bethe pointed out that this may be why the neutrino flux observed by Davis (next section) is smaller than that expected from the standard solar model.

This effect is essentially an induced neutrino oscillation in matter. The electron neutrinos, which are produced at the center of the sun, have positive effective mass due to the e-ν_e charged current interaction. During the trip through the body of the sun whose density gradually decreases, the effective mass decreases until it reaches the ν_μ mass, where ν_e converts into ν_μ. This ν_μ then escapes detection, since our detectors are only sensitive to ν_e. Since we do not know the mass of ν_μ, we do not know if this actually happens. If we could measure the whole ν_e spectrum at the surface of the earth and if the ν_μ mass, or more precisely the mass difference of ν_μ and ν_e, falls in the range of the measurable spectrum, we could determine the neutrino mass for the range much smaller than that achieved by the terrestrial measuring methods like tritium spectra or neutrino oscillation experiments.

II. Detection of Solar Neutrinos

1) Detection of Other Neutrinos

Because of the extremely small cross section for the production of any nuclear reaction by neutrons, the detection of neutrinos is several orders of magnitude more difficult than the detection of any other elementary particles. Only through missing mass and missing momentum experiments, like cloud chamber experiments of the ^{8}Li beta decay where all the energies and directions of outgoing electron and two α-particles can be measured, was the neutrino hypothesis supported.

The direct detection of neutrinos through their reactions with hadrons has, however, succeeded in a few cases. One is the case of reactor neutrinos where anti-neutrinos coming from the beta decay of the fission fragments have been detected through the $p(\bar{\nu}_e,e^+)n$ reactions. This became possible due to the very high flux of anti-neutrinos near large reactors and because of the fact that the process of the simultaneous emission of a positron and a neutron is such a specific process. This special process could be sorted out with a detector system specially designed for the purpose.

Another case of the successful detection of neutrinos is the case of accelerator neutrinos. Here, the neutrino cross section becomes considerable if its energy becomes the order of GeV. Mu-neutrinos were first observed in such an accelerator experiment.

2) Solar Neutrino Detectors

Detection of solar neutrinos depends basically on the fundamental reactions;

$$\nu_e + n \rightarrow e + p.$$

Since neutrons exist only in nuclei, the solar neutrino detectors are nuclei. Therefore, we use reactions of the form

$$\nu_e + {}_{Z}\square \rightarrow e + {}_{Z+1}\square .$$

Here, we shall discuss only this latter case. Each detector is characterized by its detection threshold, which is equal to the Q_{EC} (electron capture energy) of the nucleus ${}_{Z+1}\square$ and the neutrino cross section (the reaction rate). The cross section is proportional to the inverse beta decay rate, which is usually known as the ground state to ground state transition. The measure of the transition strength is the so-called beta decay log ft value. One must be aware of the fact that the neutrino absorption may not lead to the ground state but to an excited state. The log ft values of such excited state transitions are usually not known and often cause serious problems. Indeed, this problem seems to be the most serious problem and is a very challenging one for nuclear physicists.

The detectors must also possess favourable practical features, e.g. low enough threshold to detect the most fundamental low energy component of the solar neutrino spectrum, good isotopic abundance, low cost and availability of the nuclide and ease of chemical handling.

The detectors may be classified roughly to three categories.

i) Those which depend on the specificity of events (like the case of $\nu_e + p \rightarrow n + e^+$ as described in the case of the reactor neutrino detection). Examples are ^{115}In, ^{2}H.

ii) Radiochemical detectors. Examples are ^{37}Cl, ^{71}Ga.

iii) Geochemical Detectors. Examples are ^{205}Tl, ^{98}Mo.

In the following, three cases, ^{37}Cl, ^{71}Ga and especially ^{205}Tl will be discussed in detail.

3) The Chlorine Experiment

This is the only successful solar neutrino experiment so far and it has been running nearly twenty years guided by R. Davis. The reaction employed is

$$\nu_e + {}^{37}Cl \rightarrow {}^{37}A + e.$$

^{37}A reverts to ^{37}Cl by electron capture emitting the characteristic X-ray, with a half-life of 34 days. The greatest advantage of this reaction is the fact that a great amount of chlorine is available (also ^{37}Cl has an isotopic abundance of 25%) and the chemical procedure for extracting Argon in micro-quantities is easily done by bubbling noble gas. Unfortunately, the threshold of the reaction is too high (814 keV) to detect the main pp neutrinos (their energy is mainly below 420 keV). The problem of transition rate, or the neutrino cross section is practically well understood (this problem will be discussed separately in the last section).

As the neutrino target, 400000 litres of C_2Cl_4 was installed in a tank in a 1500m deep gold mine in South Dakota. The depth is necessary in order to suppress the cosmic ray back-ground.

The measured value is about 2 SNU (1 SNU = one solar neutrino unit = 10^{-36} events per gram per sec) and is 3 to 4 times smaller than that estimated from the standard solar model. Since the prediction of the counting rate in the standard solar model depends on T^{20}, one may think the prediction is quite model dependent. About 50 theories have already been proposed to explain this discrepancy. The recent measurement seems to show some time variation and correlation to solar activity. It has been suspected that the variation of solar magnetic field to the neutrino magnetic moments may account for this time variation.

4) The Gallium Experiment (Gallex)

^{71}Ga has many advantages as a solar neutrino detector. Especially due to its low threshold value (233 keV) this nuclide can be used to detect the pp neutrino. The isotopic abundance is 39.4% which is sufficiently large. The end product of

$$^{71}Ga(\nu_e, e)^{71}Ge$$

reaction has a favourable chemical property like chlorine. Since the produced germanium in $GaCl_3$ solution is volatile, it can be collected like in the chlorine experiment. The half-life of ^{71}Ge is 11 days which is convenient for radiochemical experiments. The log ft value of ^{71}Ge beta decay is 4.3, which is also very favourable (see the last section).

The experiment had been proposed a long time ago. Recently the Gallex-Collaboration project was financially approved and 30 tons of gallium will be acquired by the end of 1989, when the actual measurement will start. The experiment will be made in Gran Sasso, European Underground Laboratory in Italy [3]. A somewhat larger operation is going on in Baksan in the USSR.

III. The Thallium Experiment

1) Properties of ^{205}Tl as a Neutrino Detector

The threshold of the $^{205}Tl(\nu_e,e)^{205}Pb$ reaction is only $\sim$54 keV. The main transition, however, leads to the 2.3 keV excited state of ^{205}Pb. Therefore, thallium, together with gallium, is one of the most precious neutrino detectors. The half-life of ^{205}Pb is 15 million years, too long for a radiochemical detection. One must employ a geochemical method, like the case of the detection of double beta decay. What we measure is, therefore, not the present value of the neutrino flux, but an integral value of the neutrino flux over the age of minerals which we use for the detection. Here the age means the time which has elapsed since lead was separated from thallium during the formation of the ore. This large accumulation time is in a sense a great advantage. If we have a large mass spectrometer with a high collecting efficiency we should be able to perform experiments with much less target material, as in the case of radiochemical experiments. For example, if the transition rate, abundance and some other quantities are the same and if the life of the ore is longer than 15 million years, one can work with material compared to the gallium experiment with the order of $0.03y/15\times10^6 y\times x$, where x is the detection efficiency of the mass spectrometer which could be as high as 10^{-4}.

This great advantage of long accumulation time is achieved at a serious cost of uncertainty in the transition rate. The ground state transition of ^{205}Pb to $^{205}Tl(5/2^- \rightarrow 1/2^+)$ has a very high log ft value of 12.7, which is the cause of this long half-life. The neutrino capture, however, uses this transition only very rarely. The main transition is from the ^{205}Tl ground state$(1/2^+)$ to the 2.3 keV excited state in $^{205}Pb(1/2^-)$.

The log ft value of this inverse process is unknown. This uncertainty was one of the reasons why this experiment was not financed. The log ft value of the $(1/2^- - 1/2^+)$ transition is probably of the order of 5 to 6, guessing from several similar transitions in the same region of the nuclear chart. This is rather high. The problem of the transition rate (neutrino cross section) will be discussed again in the last section in more detail.

2) Freedman's Paper

A detailed and bold proposal to use ^{205}Tl as the pp neutrino detector was made in 1976 by M. Freedman and his colleagues [4]. They pointed out, after analysing the quality of a Lorandite sample from Alshar, Yugoslavia, which they found in Chicago Museum, that this mineral is a good candidate to be used as a geochemical detector for pp neutrinos, if the mineral could be obtained from a depth of several hundred meters. Thirty-four possible background reactions were checked and found to cause no serious difficulties. These workers tentatively used a log ft value for the $1/2^+ - 1/2$ transition of 5.3 and assumed that the mineral is 10 million years old and estimated the event rate of 430 SNU. For the detection of ^{205}Pb, a double focusing spectrometer was proposed. This proposal, however, was not materialized.

3) Detection of ^{205}Pb with Accelerator Mass Spectrometry

A few years ago, we started some preliminary experiments to detect ^{205}Pb by accelerator mass spectrometry(AMS). This method of extremely high sensitivity to detect abundances of the order of 10^{-10} to 10^{-15} was essentially started by Litherland, Gove and Purser in the same year that Freedman wrote about the thallium experiment. Here they emphasized that ^{14}C can be better measured by counting directly the number of nuclei with a mass spectrometer than by measuring its slow beta decay.

The merit of AMS is that the nuclei to be measured are accelerated to such a high energy that they can be counted one by one microscopically, not as a current or a blackening of an emulsion, where macroscopic quantities are measured.

AMS of light elements can be done with very small accelerators but in order to obtain a good identification of heavy nuclei, use of larger accelerators is more favourable. Since our group has had several years experience in AMS for relatively heavy elements (chlorine 36) with very long path lengths, we tried to identify ^{205}Pb in ^{205}Tl at the large heavy-ion accelerator UNILAC at GSI, Darmstadt [5].

If we take Freedman's estimate of capture rate, 430 SNU; age of the mineral 10×10^{6}y and lead concentration of 3 ppm, we get $^{205}Pb/^{205}Tl\simeq10^{-19}$, $^{205}Pb/Pb\simeq0.3\times10^{-13}$. The latter is a good value to be measured by AMS. The result of our test experiment at GSI turned out to be

a) Suppression of neighbouring element 10^{3}

b) Suppression of neighbouring masses 10^{16}

c) Chemical composition of the sample required is $Tl/Pb\sim10^{-10}$.

The present ion sources are not designed for AMS, but if we obtain an ion source which yields total detection efficiency of 10^{-4}, then we need for 100 events 10^{6} ^{205}Pb atoms. These will be contained in 11 mg of Pb extracted from 7.6 kg lorandite. This can be collected from a few tons of ore, which is a reasonable quantity.

4) Feasibility Study of the Thallium Experiment

Since, as discussed above, AMS seems to offer a good possibility of detecting ^{205}Pb produced in the lorandite from Alshar, we have begun a project to study the quality of the lorandite from Alshar in more detail, together with Yugoslavian geologists and physicists and also with some other laboratories, mainly in Germany. We have organized two workshops for this purpose, one on 23,24/9/1985 in Munich and the other on 23-27/9/1986 [6] in Dubrovnik.

What should be done is to find out if lorandite really suited for the thallium experiment can be obtained, and to measure different parameters, like life time of the minerals, different impurities in the ore and so on. The proceedings of the Dubrovnik workshop will appear as a special issue of Nuclear Instrument and Methods.

The most serious problem is that of obtaining good lorandite from sufficient depth. The uranium and thorium content in the rocks nearby seems to be satisfactorily small. Therefore, most of the dangerous background reactions come mainly from cosmic rays. These diminish sharply as the depth increases. Feasibility studies are already being made by several groups including our own. We found that AMS is very useful for such studies. We are now using ^{36}Cl, ^{41}Ca, ^{53}Mn for such studies. The chemical processing of the mineral is also quite a complicated problem.

IV. Transition Rate problems

1) Radiochemical Yields of Neutrino Reactions

If we have a large number of radiochemical and geochemical neutrino detectors with different thresholds, we can obtain $N(\varepsilon_\nu)$, the complete solar neutrino (ν_e) spectrum, given the response function $\sigma(\varepsilon_\nu, \varepsilon_e)$ of each detector

$$A_j = \int_0^{\varepsilon_p} N(\varepsilon_\nu)\, \sigma_j(\varepsilon_\nu, \varepsilon_e) d\, \varepsilon_\nu$$

where ε_e is the excitation energy of the residual nucleus and ε_p is the particle threshold. The latter limit may be important if (ν_e, e) reactions compete with (ν_e, e n) reactions (absorption of neutrino with such a high energy that one further neutron is emitted) at higher excitation energy. $\varepsilon_\nu - \varepsilon_e$ is the threshold energy.

Even if $\sigma_j(\varepsilon_\nu, \varepsilon_e)$ is known, the measurement of $N(\varepsilon_\nu)$ is practically impossible since the number of available detectors (j) is extremely limited (at most ten). The $\sigma_j(\varepsilon_\nu, \varepsilon_e)$ is in some cases (for example for ^{37}Cl) easily deduced. But in other cases it cannot be readily obtained. In some cases, the main part is only one transition and this could be the ground state transition. In such a case, we can deduce $\sigma_j(\varepsilon_\nu, \varepsilon_e)$ roughly from the log ft of the inverse beta decay.

Usually, the log ft value for the beta decay of the excited state of the final nucleus to the target ground state is unknown because the life time of the gamma decay of the excited state is much shorter than the beta decay. This could sometimes be a serious problem.

For such cases, it was suggested long ago that the beta decay matrix element could be obtained from the forward cross section of the (p,n) reactions with several hundred MeV protons.

We shall discuss this problem for three practical cases of ^{37}Cl, ^{71}Ga and ^{205}Tl detectors.

2) ^{37}Cl Detector

This is a very good case since for all states which can be obtained by rather high energy transitions, the transition probabilities to the excited states (with low log ft values) are known. This is because of the mirror decay of the ^{37}A to ^{37}Cl, the decay of ^{37}Ca to ^{37}K is measured. We can therefore deduce the log ft value of all the important transitions caused by boron and CNO neutrinos.

3) ^{71}Ga Detector

As mentioned before, the ground state transition is rather fast (log ft = 4.3). Therefore, the disturbances from other states are relatively small. Here (p,n) data exist and support the above-mentioned contribution. According to the standard solar model the boron neutrino contribution to the higher excited state is about 10%.

4) ^{205}Tl Detector

Here, the transition rate problem has been a serious handicap and became the reason for rejecting the project. However, very important progress has been made quite recently due to improvements in nuclear spectroscopy and due to an ingenious idea.

As briefly discussed before, the most serious problem is the fact that the log ft value of the main transition for detecting the pp neutrinos has not been measured and the prospects were thought to be hopeless. This is the neutrino capture by the ground state of ^{205}Tl($1/2^+$) to the 2.3 keV first excited state of ^{205}Pb($1/2^-$). The latter then decays to the ground state of ^{205}Pb($1/2^-$) with an E2 transition. The beta branch to the ground state of ^{205}Tl can never compete with this gamma transition. The possible use of the (p,n) or the (^{3}He,t) reactions has not been investigated seriously. Anyway, the transition in question is not a well-studied and favoured transition. This possibility of using such reaction with good calibration is worth studying.

Theoretical estimation is another way to approach the problem. There are nearly ten transitions of $1/2^+$-$1/2^-$ near lead. Old calculations by Kurath, however, give rather a large variation in the estimate according to the wave function used. This is a rather discouraging result. The spectroscopy of the lead region, however, has made significant progress in the last few years. Carefully calibrated wave functions with adequate data will, according to the opinion of specialists, yield reliable values for use in the estimation of the neutrino flux [7].

Recently, P. Kienle has proposed an ingenious idea to measure this transition probability experimentally [8]. The idea is to use the possibility of bound state beta decay. The binding energy of the K electron in Tl is higher than Q_{EC} of ^{205}Pb. Therefore, if one takes out all the electrons from a thallium atom, the thallium 205 nucleus can change into a lead 205 nucleus by emitting an electron to the bound state. Experimentally, one accelerates ^{205}Tl to such high energies that by collision with material all the electrons will be stripped off. This naked ^{205}Tl will be stopped in a storage ring and one awaits the bound state decay into ^{205}Pb. Freedman calculated in detail this half-life and found it to be of a measurable size [9].

There is another problem in the case of ^{205}Tl where the main transition is not as fast as in the case of ^{71}Ga. There is possible background due to higher energy neutrinos, especially the boron neutrinos. In the case of heavy nuclei, high energy transitions of the favoured type become non-negligible due to neutron excess. The favoured transitions due to the neutron excess are not blocked by the Pauli principle. A simple calculation shows that this amounts to as much as 70 SNU for boron neutrinos if their intensity is taken from the standard solar model [10]. More detailed calculations are necessary.

5) Other cases

The problem of transition probability is, as shown before, quite an individual problem especially for low lying excited states. There are already several other detectors under consideration, but for example in the case of ^{98}Mo, the most important transtition to the lowest 1^+ state in ^{98}Tc has not yet been accurately measured. It seems to me that there is still more to be done using the well-developed techniques of modern nuclear physics.

References

1. J.N. Bahcall, W.F. Huebner, S.H. Lubow, P.D. Parker and R.K. Ulrich, Rev. Mod. Phys. 54, 767 (1982).
2. S. Mikheyev and A. Smirnov, Yad. Fiz. 42, 1441 (1985).
3. T. Kirsten, to appear in the Proceedings of VIth Moriond Workshop on Massive Neutrinos in Particle Physics and Astrophysics, Tignes, Savoie, France 25 Jan.-1 Feb., 1986.
4. M.S. Freedman, C.M. Stevens, E.P. Horovitz, L.H. Fuchs, J.S. Lamer, L.S. Goodman, W.J. Childs and J. Hessler, Science 193, 1117 (1976).
5. M. Ernst, G. Korschinek, P. Kubik, W. Mayer, H. Morinaga, E. Nolte, U. Ratzinger, W. Henning, W. Kutschera, M. Müller and D. Schüll, Nucl. Instr. and Meth. B5, 426 (1985).
6. Proceeding will appear in Nucl. Instr. and Meth.
7. Private communication from J. Blomquist and I. Hamamoto; see also K. Ogawa in ref.[6].
8. P.Kienle, in ref.[6].
9. M.S. Freedman, in ref.[6].
10. H. Morinaga, in Current Problems in Nuclear Physics, Hellenic Physical Society, Conference Series, Vol.1, 141 (1986).

What Are Nuclei Made Of?

K. Yazaki

Department of Physics, Faculty of Science,
University of Tokyo, Tokyo 113, Japan

The nucleus is usually considered as a composite system of nucleons with small admixtures of mesons and isobars. These constituents of the nucleus are now considered to be also composite systems of quarks and gluons. The models of the internal structure of the nucleon are discussed in the context of their impacts on nuclear physics. The quark cluster model is taken in particular to study the quark effects on nuclear force and nuclear response functions.

§1 Introduction

The nucleus manifests its different aspects of compositeness, depending on how it is experimentally probed. Most of the low energy phenomena reveal only the aspect of a composite system of nucleons. At higher energies where mesons and isobars can actually be produced, we study the features of a hadron many-body system. The quark and gluon degrees of freedom may show up in very high energy or short distance phenomena. The answer to the question "What are nuclei made of ?" thus depends on which aspect we are interested in. The question can be converted to "What variables are the most convenient in describing the phenomena of interest ?" and "What are the more fundamental variables ?"

Nuclear physics has a long history of searching for such variables even within the framework of the nucleus as a many-nucleon system. The only exactly solvable systems for given nuclear forces are two nucleon systems (nucleon-nucleon scattering and the deuteron) and the three-nucleon bound state (the triton). One thus needs some models for the description of more complex systems. The shell model [1] takes account of only the variables of valence nucleons; the cluster model [2] treats only the relative motions among clusters; the collective model [3] uses the variables describing the surface deformation, and the interacting boson model [4] chooses a few kinds of bosons to describe the complicated spectra.

The Hamiltonian for a many-nucleon system is given by

$$H = \sum_i K_i + \sum_{i<j} V_{ij} + \sum_{i<j<k} V_{ijk} + \cdots \qquad (1)$$

and the interaction with an external field is described by a current of the form

$$J = \sum_i J_i + \sum_{i<j} J_{ij} + \cdots \qquad (2)$$

The one-body parts, K_i and J_i, are determined by the properties of a single nucleon, the two-body parts, V_{ij} and J_{ij}, are, in principle, determined by the analyses of two-nucleon systems, and so forth. Such a description is complete in the energy region where mesons and isobars are not produced. In fact, most of the nuclear phenomena have been successfully described by truncating the sums in eqs. (1) and (2) after the two-body terms. (There is an indication from the calculation [5] of the three-nucleon system that one needs the three-body force V_{ijk} to explain the binding energy of ^{3}H.)

In order to treat higher energy phenomena, one has to go beyond the above description. Such a step is also necessary if one wants to understand K_i, V_{ij}, J_i, J_{ij} etc. at a more fundamental level. The explicit introduction of mesons and isobars is certainly required for the description of the processes where these particles are actually produced. They are also necessary to understand the nuclear force, V_{ij}, and the exchange current, J_{ij}. The meson theory of nuclear force starts from Yukawa's paper [6] and has been refined to give almost a complete description [7] of deuteron properties and nucleon-nucleon scattering observables. The meson exchange currents have been also confirmed in the deuteron photo-disintegration [8], the electromagnetic form factors of ^{3}H and ^{3}He [9], and nuclear magnetic moments [10]. There are thus clear indications that hadrons other than the nucleons play important roles in nuclear physics.

What about quarks and gluons, then ? Since they are considered to be absolutely confined and the ultimately observable objects are hadrons, the description in terms of hadrons should in principle be complete. It will, however, become extremely complicated at very high energies where many hadrons can be produced. This is the case, for example, in lepton deep inelastic scatterings, while the quark-parton model [11] gives a very simple and intuitive description of the processes. The quark-gluon description will also be indispensable for heavy-ion collisions at extremely high energies where the quark-gluon plasma [12] is expected to be temporarily formed. The lepton pair production in hadron-hadron collisions at very high energies (the Drell-Yan process) [13] is also considered to reveal the quark structure of the hadrons. There are thus many phenomena at high energies for which the quark-gluon description is simpler and more natural than other descriptions although it may not be absolutely necessary.

Are there any phenomena at lower energies which are described by the quark variables in a simple and natural way ? We will show in the next section that the internal structure of the nucleon might be such a candidate. Among the various models of the nucleon, we take the quark cluster model and see how we can apply the model to the description of nuclear phenomena. The model is briefly explained in §3. The applications to the nuclear force and the nuclear response functions are described in §4 and §5, respectively. Concluding remarks are given in §6.

§2 Models of the Nucleon

The importance of the nucleon internal structure was first suggested by its magnetic moment. If it were an elementary Dirac particle, the proton (neutron) magnetic moment would be 1.0 (0.0) μ_N (μ_N: nuclear magneton). The observed value is 2.79 (-1.91) μ_N for the proton (neutron). Further evidence for the internal structure has been provided by electron scattering experiments. The spatial extension of the nucleon has been studied by the measurement of its electromagnetic form factors while the internal excitations have been revealed by the measurement of the response (structure) functions. For large energy and momentum transfers, the structure functions exhibit the scaling behavior and have been successfully described by the quark-parton model. At lower excitation energies, they show the resonance peaks corresponding to the isobars. The phenomenological models of the nucleon are usually constructed so as to describe these isobar levels together with the static properties (form factors).

Among such models are the non-relativistic quark model or the quark cluster model [14], the MIT bag model [15] and its chiral versions (the chiral bag [16] and the cloudy bag [17]), Skyrme's model [18] and its versions with vector mesons [19], the soliton bag model [20] and its chiral versions [21] and the color dielectric model [22] and its chiral versions [23]. All of them describe the static properties reasonably well but the excitation spectra have been successfully described only by the quark cluster model. We therefore take this model here and apply it to the description of nuclear phenomena. The model has another advantage, that multi-nucleon systems and the single nucleon can be treated on the same footing. It has, however, a disadvantage that the chiral symmetry cannot be easily incorporated. In order to refine the model to give a more quantitative description of the static properties and the excited levels including their decays, one has to couple mesons, at least pions, with the quark cluster. There have been attempts [24] to introduce mesons as quark-antiquark excitations. I personally prefer the other view [25] of regarding the pion as the Goldstone particle associated with the chiral symmetry which can be approximately treated as an elementary particle and can be directly coupled with quarks. In any case, the model makes sense only if the coupling is weak enough to be included in a perturbative way. It should also be noted here that the constituent quarks appearing in the model are considerably different from the current quarks which are relevant in the quark-parton model and the perturbative QCD. One has to clarify the relation between the above two kinds of quarks before one can get any foundation of the model based on QCD.

§3 Quark Cluster Model

This is a model of hadrons with the valence quark degrees of freedom, their equation of motion being described by a non-relativistic Hamiltonian. We restrict ourselves here to systems of a baryon or baryons.

The wave function for a single baryon consisting of three quarks, 1, 2 and 3, has the form

$$\psi_B(1,\ 2,\ 3) = [\phi(\vec{r}_{1,\ 23},\ \vec{r}_{23})\ \ S(S_1,\ S_2,\ S_3)]_B\ C(C_1,\ C_2,\ C_3) \tag{3}$$

where C denotes the totally antisymmetric color singlet state, and the orbital part, ϕ, and the spin-flavor part, S, are combined to give a totally symmetric state specifying the baryon, B. $\vec{r}_{1,\ 23}$ denotes the relative coordinate between the quark 1 and the center-of-mass for the quarks 2 and 3. Similar notations will be used later. For a nucleon, the orbital part is usually taken to be a Gaussian which corresponds to the $(0S)^3$ configuration in the harmonic oscillator model and its size parameter, b, which coincides with the r.m.s. radius of the nucleon, is chosen to be in the range of 0.5 ∿ 0.8 fm.

A system of two baryons consisting of six quarks, 1, 2, ⋯, 6, is described by a wave function of the form

$$\psi_2(1,\ 2,\ \cdots,\ 6) = \sum_{B,B'} \mathcal{A}[\psi_B(1,\ 2,\ 3)\psi_{B'}(4,\ 5,\ 6)\chi_{BB'}(\vec{r}_{123,\ 456})], \tag{4}$$

where $\chi_{BB'}$ is the wave function for the relative motion between baryons, B and B', and $\mathcal{A}$ is the antisymmetrizer for quarks. The wave function for a system of more baryons can be constructed analogously. A three-baryon system, for example, is described by a wave function of the form

$$\psi_3 = \sum_{B, B', B''} \mathcal{A}[\psi_B\psi_{B'}\psi_{B''}\chi_{B, B', B''}], \tag{5}$$

where $\chi_{B, B', B''}$ now contains two relative coordinates among three baryons, B, B' and B''.

The Hamiltonian for quarks is given by

$$H = \sum_i (m_i + \frac{\vec{p}_i^{\,2}}{2m_i}) + \sum_{i<j} V_{ij}^{OGEP} + V^{CONF}, \tag{6}$$

where the mass, m_i, is taken to be about 300 MeV for u and d quarks. V^{OGEP} is the one-gluon exchange potential and V^{CONF} describes the confinement. The conventional choice is the two-body model, i.e.

$$V^{CONF} = - a \sum_{i<j} (\lambda_i \cdot \lambda_j) r_{ij}^{\,2}, \tag{7}$$

where λ is a set of color SU(3) generators. This model is known to have a problem of the long range color van der Waals force [26] although, in any practical calculation, the truncation of the wave function gives an automatic screening of the force and covers the potential difficulty. Another possible choice is the flip-flop model [27] which has recently been proposed with the aim of avoiding the above problem. It is identical to the two-body model for a single baryon (three quark) system but gives an essentially many-body interaction for multi-baryon systems. The details of the model have been described in ref. [27].

§4 Nuclear Force

The meson exchange model of the nuclear force has been extensively studied since the famous paper [6] by Yukawa in 1935. The longest range part due to one pion exchange has now been well established. The medium range part has also been fairly well clarified and is considered to be due to the exchange of two pions, and heavier mesons i.e. η ρ and ω. The short range part, on the other hand, is usually described in a phenomenological way by assuming a hard core, or a soft core or a velocity-dependent core. The quark cluster model has been applied mainly to the description of this part.

One assumes a wave function of the form (4) with the sum over B and B' being restricted to two-nucleon channels, i.e.

$$\psi_2 = \mathcal{A}[\psi_N\psi_N\chi]. \tag{8}$$

A variational principle for χ gives

$$\langle\psi_N\psi_N|(E - H)|\psi_2\rangle = 0 \tag{9}$$

which can be written as an integral equation for χ, i.e.

$$\int d\vec{r}'\{EN(\vec{r}, \vec{r}') - H(\vec{r}, \vec{r}')\}\chi(\vec{r}') = 0 \tag{10}$$

where the norm kernel, $N(\vec{r}, \vec{r}')$, and the Hamiltonian kernel, $H(\vec{r}, \vec{r}')$, can be calculated with the internal wave function, ψ_N, for the nucleon and the Hamiltonian (6). In order to take account of the meson exchange effect, which cannot be included in the present version of the quark cluster model, we supplement the Hamiltonian kernel with a term of the form

$$\delta H(\vec{r}, \vec{r}') = \int d\vec{r}''\, N^{1/2}(\vec{r}, \vec{r}'')v^{EMEP}(\vec{r}'')N^{1/2}(\vec{r}'', \vec{r}') \tag{11}$$

where v^{EMEP} consists of the one pion exchange tail and a medium range attraction which is taken to be of the Gaussian form with the depth parameter, v_0^C, and the range parameter, α_0^C. See ref. [28] for the details of the parametrization.

Figure 1 compares the calculated $^1S_0$ phase shifts with the observed ones. The calculation was made for three values of the size parameter, b, i.e. (I) b = 0.5 fm, (II) b = 0.6 fm and (III) b = 0.7 fm, with and without effective meson exchange potential, v^{EMEP}. In the calculation with v^{EMEP}, the para-

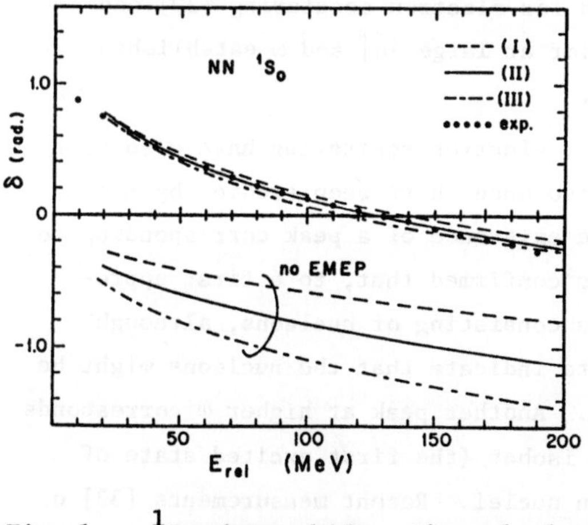

Fig. 1. $^1S_0$ phase shift. The calculated ones for three values of the size parameter, b, i.e. (I) b = 0.5 fm, (II) b = 0.6 fm and (III) b = 0.7 fm, with and without the effective meson exchange potential, are compared with the observed one.

meters, v_0^C and α_0^C, were adjusted to give the correct low energy parameters.

The two-body model was used for the confinement. One sees that the quark part gives short-range repulsion with the strength depending on the size parameter, b, and supplemented with an appropriate effective meson exchange potential, reproduces the experimental data fairly well. The flip-flop model of confinement gives similar results for the short-range repulsion. It has more parameters than the two-body model and offers the possibility of explaining an appreciable part of the medium range attraction by a suitable choice of these parameters [29]. Such a choice has also been used to calculate the ^3P phase shifts and is found to explain the major part of the nucleon-nucleon spin-orbit interaction without sacrificing the single baryon spectrum.

§5 Nuclear Response Functions

The excitation spectrum and the constituents of a system are most effectively studied through its response functions (structure functions in the terminology of particle physics) for various probes such as electron, muon, neutrino, photon and hadrons. The response function is defined by

$$R(\vec{q}, \omega) \equiv \sum_f \delta(\omega - E_f + E_i)\,|\langle f|\hat{O}(\vec{q}, \omega)|i\rangle|^2 \tag{12}$$

where $\vec{q}$ and ω are the momentum and the energy transferred to the system, respectively, $|i\rangle(|f\rangle)$ is the initial (final) state of the system with the energy E_i (E_f) and $\hat{O}(q, \omega)$ is the operator describing the interaction with the projectile. As was mentioned in the introduction, the nucleon response (structure) functions have been measured for electron scattering [30] and neutrino reactions [31] and their behavior at large $|\vec{q}|$ and ω established the quarks as the ultimate constituents.

The response functions of nuclei for electron scattering have also been measured extensively [32]. The giant resonances have been studied by the response function at low $|\vec{q}|$ and ω. The existence of a peak corresponding to the single-nucleon knock-out process has confirmed that, to a first approximation, a nucleus can be considered as consisting of nucleons, although a closer examination of the peak seems to indicate that the nucleons might be slightly modified in the nuclear medium. Another peak at higher ω corresponds to the quasi-free production of the Δ - isobar (the first excited state of the nucleon) and reveals its behavior in nuclei. Recent measurements [33] of the response functions at large $|\vec{q}|$ and ω showed that the quark distributions in nuclei are appreciably different from that of a nucleon (the EMC effect).

Hundreds of theoretical papers concerning this effect have been published. One of the speculations is that the confinement scale is modified in the nuclear medium [34]. A more conservative explanation is due to a kinematical effect of nuclear binding [35]. There is also an attempt [36] to make a connection between these two apparently different explanations. We will not discuss this subject further since it is certainly not in the regime where the quark cluster model is supposed to make sense.

We use here simplified wave functions for very light nuclei, i.e. ^3H (or ^3He) and ^4He, within the framework of the quark cluster model, to see how the quark effect manifests itself in the nucleon knockout response functions [37]. The wave function for ^4He is taken to be

$$\Psi_4 = \mathcal{A}\,[\psi_N\psi_N\psi_N\psi_N\chi] \tag{13}$$

where the nucleon internal wave function, ψ_N, is assumed to be the same as was explained in §3, and the spin-isospin parts of the four nucleons are combined to give the correct quantum numbers for ^4He. The wave function, χ, describing the relative motion among the four clusters (nucleons) is assumed to be a Gaussian with the size parameter, b_N. If we take b_N to be $b/\sqrt{3}$ (b is the size parameter of the internal motion), the wave function, Ψ_4, becomes that of the harmonic oscillator quark shell model with the configuration $(0S)^{12}$. We therefore introduce the ratio

$$\beta \equiv b/(\sqrt{3}b_N) \tag{14}$$

as a measure of the overlap among nucleon clusters, $\beta = 0$ and 1 corresponding to the point nucleon and the quark shell model limit, respectively. Figure 2 shows the expectation value of the antisymmetrizer, $\mathcal{A}$, defined by

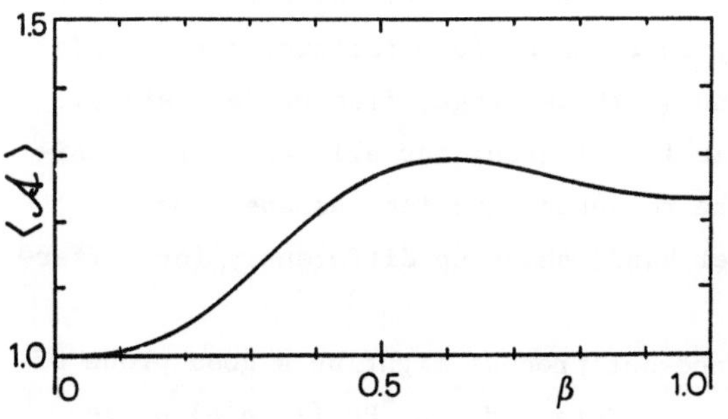

Fig. 2. The expectation value of the antisymmetrizer, $\mathcal{A}$, for ^4He as a function of the parameter, β.

$$< \mathcal{A} > \equiv \langle \psi_N\psi_N\psi_N\psi_N\chi | \mathcal{A} | \psi_N\psi_N\psi_N\psi_N\chi \rangle \tag{15}$$

as a function of β. If the parameter b_N is adjusted for each value of b so that Ψ_4 gives the observed r.m.s. radius of ^{4}He, the range of b = 0.5 - 0.8 fm corresponds to $\beta = 0.2 \sim 0.3$. One sees that the expectation value $< \mathcal{A} >$ deviates appreciably from 1 in this region, which indicates that the Pauli effect at the quark level can be important in the plausible range of nucleon size. Taking a similar wave function, Ψ_3, for ^{3}H (or ^{3}He), we have calculated the nucleon knockout response function, defined by

$$R^{(\ell)}(q, \omega) = \int d\vec{k}_N d\vec{k}_3\ \delta(\vec{k}_N + \vec{k}_3 - \vec{q})\delta(\frac{k_N^2}{2m_N} + \frac{k_3^2}{6m_N} - \omega)$$
$$\times |\langle \psi_N(\vec{k}_N), \Psi_3(\vec{k}_3) | \sum_i O_i^{(\ell)} e^{i\vec{q}\vec{r}_i} | \Psi_4 \rangle|^2 \tag{16}$$

where $\vec{k}_N$ and $\vec{k}_3$ are the momenta of the nucleon and the three-nucleon system (^{3}H of ^{3}He), respectively, $\sum_i$ is the sum over 12 quarks and the operator $O^{(\ell)}$ specifies the type of transition, i.e.

$$\begin{aligned} O^{(1)} &= 1 \text{ (isoscalar spin-independent)}, \\ O^{(2)} &= \sigma_3 \text{(isoscalar spin-dependent)}, \\ O^{(3)} &= \tau_3 \text{(isovector spin-independent)}, \\ O^{(4)} &= \sigma_3\tau_3 \text{ (isovector spin-dependent)}. \end{aligned} \tag{17}$$

Because of the spin-isospin symmetry in the present treatment, the two response functions, $R^{(2)}$ and $R^{(3)}$, are identical. Figure 3 shows the response functions, $R^{(\ell)}$ (ℓ = 1, 2, 4), at q = 3 fm^{-1} calculated with (solid lines) and without (dashed lines) quark exchange effect, for three values of b, i.e. (1) b = 0.8 fm, (free nucleon size), (2) b = 0.88 fm (10% extended nucleon) and (3) $\beta = 1$ (quark shell model limit). The conventional calculation, which takes account of the nucleon internal structure only in terms of form factors, corresponds to the dashed lines of the case (1) (no quark exchange, free nucleon size). The effect of changing the nucleon size is universal for all types of transitions and the increase of b reduces the response functions as one expects. The quark exchange effect, on the other hand, shows up differently for different types of transitions.

One thus sees that the nucleon-knockout process might be a good probe of the quark effects in nuclei. Experimental data [38] on ^{4}He (e, e'p)^{3}H are available at $q \simeq 2.1$ fm^{-1}. The data are roughly consistent with the conven-

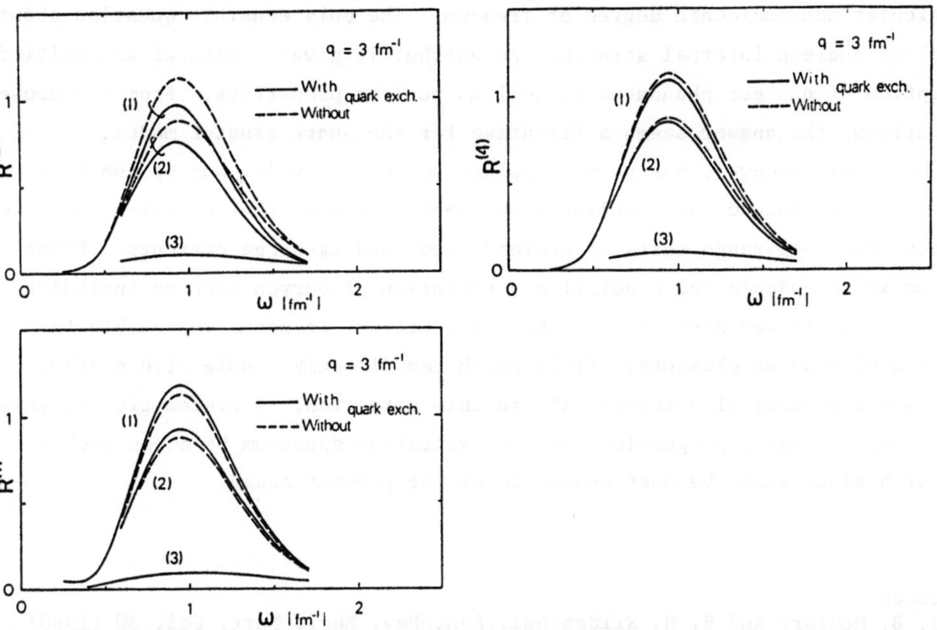

Fig. 3. Nucleon knockout response functions, $R^{(\ell)}(q, \omega)$, at $q = 3\ \mathrm{fm}^{-1}$ calculated for three cases, i.e. (1) $b = 0.8$ fm (free nucleon size), (2) $b = 0.88$ fm (10% extended nucleon) and (3) $\beta = 1$ (quark shell model limit)

tional description of ^{4}He and therefore exclude the quark shell model limit (case (3)). For a more quantitative comparison, one needs refined data with longitudinal and transverse parts separated as well as a detailed calculation based on the conventional description. The reaction ^{4}He (e, e'n) ^{3}He seems particularly interesting since the isoscalar ($\ell = 1$) and isovector ($\ell = 3$) transitions interfere in such a way that only the quark exchange effect survives.

§6 Concluding Remarks

We have discussed here implications of the quark cluster model for nucleons and nuclei. The model assumes that a nucleon is a three-quark cluster and a nucleus is an assembly of these clusters interacting with each other through the interchange of quarks. It is known to be successful in describing the static properties and the excitation spectrum of a nucleon. We saw that it also explains the short-range part of nuclear force and predicts some observable effects of the quark exchange in nuclear response functions. It should be emphasized, however, that all these effects can be expressed in the form of two-body interactions and exchange currents for nucleons. As was mentioned in the introduction, the description in terms of nucleon variables is complete for low energy phenomena and it is impossible to ascribe any observable effects to a particular non-nucleonic degree of freedom. The only sensible question about a model of nucleon internal structure is whether it gives a natural and unified description of nuclear phenomena as well as nucleon properties. From the above observations, the answer seems affirmative for the quark cluster model.

The model, however, has several points to be improved. One of the most urgent is to introduce the coupling with pions. We know that one-pion exchange dominates the long-range parts of nuclear force and exchange currents. Pions are also indispensable for a detailed description of baryon spectra including their decays. As was mentioned in the introduction, it seems reasonable to treat the pion as an elementary field which can directly couple with quarks. There have been several attempts [25] in this direction. A systematic analysis of the nucleon static properties and the excitation spectrum based on such a model with pions would be most desirable at the present stage.

References

1. J. B. McGrory and B. H. Wildenthal, Ann. Rev. Nucl. Part. Sci. 30 (1980) 383.
2. K. Ikeda, H. Horiuchi and S. Saito, Prog. Theor. Phys. Suppl. No. 68 (1980) 1.
3. A. Bohr and B. R. Mottelson, Nuclear Structure vol. II, Benjamin (1975).
4. A. Arima and F. Iachello, Ann. Phys. (N. Y.) 123 (1979) 468.
5. T. Sasakawa, Proceedings of the 11th Int. Conf. on Few Body Systems in Particle and Nuclear Physics, Tokyo and Sendai, 1986. ed. T. Sasakawa, K. Nishimura, S. Oryu and S. Ishikawa, to be published.
6. H. Yukawa, Proc. Phys. Math. Soc. Japan, 17 (1935) 48.
7. M. Lacombe et al., Phys. Rev. C21 (1980) 861. K. Holinde, Phys. Rep. 68C (1981) 121.
8. D. O. Riska and G. E. Brown, Phys. Lett. 38B (1972) 193.
9. P. Sauer, Proceedings of the 11th Int. Conf. on Few Body Systems in Particle and Nuclear Physics, Tokyo and Sendai, 1986 ed. T. Sasakawa, K. Nishimura, S. Oryu and S. Ishikawa, to be published.
10. T. Yamazaki, in Mesons in nuclei, ed. D. H. Wilkinson and M. Rho (North Holland, Amsterdam, 1979) 651.
11. See e. g. T. D. Lee, Particle Physics and Introduction to Field Theory, Harwood Academic Publishers, 1981, p. 702.
12. See e. g. L. van Hove, Proc. 5th Int. Conf. on Ultra Relativistic Nucleus-Nucleus Collisions, Asilomar, 1986. ed. L. S. Schroeder and M. Gyulassy, Nucl. Phys. A461 (1987) 3C.
13. See e. g. G. T. Garvey, Proc. Int. Conf. on a European Hadron Facility, Mainz, 1986. Nucl. Phys. B279 (1987) 221.
14. N. Isgur and G. Karl, Phys. Rev. D18 (1978) 4187, D19 (1979) 2653, D20 (1979) 1191.
15. A. Chodos, R. L. Jaffe, K. Johnson, C. B. Thorn and V. F. Weisskopf, Phys. Rev. D9 (1974) 3471; P. Hasenfratz and J. Kuti, Phys. Rep. 40 (1978) 75.
16. A. Chodos and C. B. Thorn, Phys. Rev. D12 (1975) 2733; G. E. Brown, M. Rho and V. Vento, Phys. Lett. 84B (1979) 83.
17. S. Theberge, A. W. Thomas and G. A. Miller, Phys. Rev. D22 (1980) 2838.

18. T. H. R Skyrme, Proc. Roy. Soc. A260 (1961) 127; G. S. Adkins, C. R. Nappi and E. Witten, Nucl. Phys. B223 (1983) 433; I. Zahed and G. E. Brown, Phys. Rep. 142 (1986) 1.
19. T. Fujiwara, Y. Igarashi, A. Kobayashi, H. Otsu, T. Sato and S. Sawada, Prog. Theor. Phys. 74 (1985) 128; G. S. Adkins, Phys. Rev. D33 (1986) 193.
20. R. Friedberg and T. D. Lee, Phys. Rev. D18 (1978) 2623; R. Goldflam and L. Wilets, Phys. Rev. C25 (1982) 1951.
21. M. C. Birse and M. K. Banerjee, Phys. Rev. D31 (1985) 118; S. Ohta and R. Seki, Nucl. Phys. A458 (1986) 669.
22. A. Schuh and H. J. Pirner, Phys. Lett. 173B (1986) 19. A. G. Williams, L. R. Dodd and A. W. Thomas, Phys. Lett. 176B (1986) 158.
23. I. Duck, Phys. Rev. D34 (1986) 1493. H. Kitagawa, Master Thesis, Univ. Tokyo, 1987.
24. Y. W. Yu and Z. Y. Zhang, Nucl. Phys. A426 (1984) 557. Y. Fujiwara and K. T. Hecht, Nucl. Phys. A444 (1985) 541, A451 (1986) 625.
25. W. Weise, Progress in particle and nuclear physics. (Erice, 1983) ed. D. H. Wilkinson, p. 123; K. Shimizu, Phys. Lett. 148B (1984) 418.
26. S. Matsuyama and H. Miyazawa, Prog. Theor. Phys. 61 (1979) 942; O. W. Greenberg and H. J. Lipkin, Nucl. Phys. A370 (1981) 349.
27. F. Lenz, T. Londergan, E. Moniz, R. Rosenfelder, M. Stingl and K. Yazaki, Ann. Phys. (USA) 170 (1986) 65; Y. Koike, Nucl. Phys. A454 (1986) 509.
28. M. Oka and K. Yazaki, Quarks and Nuclei, ed. W. Weise, World Scientific, 1984, ch. 6, p. 490.
29. Y. Koike and K. Yazaki, Phys. Lett. B179 (1986) 332.
30. S. Stein et al., Phys. Rev. D12 (1975) 1884.
31. See. e. g. Review of Particle Properties, Phys. Lett. 170B (1986) 79.
32. P. Barreau et al., Nucl. Phys. A402 (1983) 515; J. S. O'Connel et al., Phys. Rev. Lett. 53 (1984) 1627.
33. J. Aubert et al., Phys. Lett. B123 (1983) 275; A. Bodek, et al., Phys. Rev. Lett. 50 (1983) 1431; R. G. Arnold et al., Phys. Rev. Lett. 52 (1984) 727.
34. R. L. Jaffe, F. E. Close, R. G. Roberts and G. G. Ross, Phys. Lett. B134 (1984) 449; Phys. Rev. D31 (1985) 1004.
35. S. V. Akulinichev, S. K. Kulagin and C. M. Vagradov, Phys. Lett. B158 (1985) 485.
36. F. E. Close, R. G. Roberts and G. G. Ross, Phys. Lett. B168 (1986) 400 Phys. Lett. B183 (1987) 101.
37. S. Takeuchi, K. Shimizu and K. Yazaki, Nucl. Phys. A449 (1986) 617.
38. V. A. Gol'tshtein et al., Sov. J. Nucl. Phys. 19 (1974) 369.

The Quantum Hall Effect

T. Ando

Institute for Solid State Physics, University of Tokyo,
7-22-1 Roppongi, Minato-ku, Tokyo 106, Japan

A recent development in understanding the quantized Hall effect is reviewed with emphasis on effects of electron localization in strong magnetic fields. A brief description is also given of the ground state and the fractional quantum Hall effect.

1. INTRODUCTION

In a two-dimensional system in a strong magnetic field, the orbital motion of electrons is completely quantized and the energy spectrum comprises discrete Landau levels [1]. The transport properties of this system with such a singular density of states is of fundamental importance. One of the exciting phenomena is the quantum Hall effect. This effect has been proposed as a method of high precision measurement of the universal quantity e^2/h and can also be used as a supreme resistance standard [2].

The quantized Hall effect requires the presence of both localized and extended states in strong magnetic fields and therefore is closely related to the Anderson localization due to random potential fluctuations [3]. The localization problem itself is quite interesting in its own right. In this paper, a brief review is given of the theory of the quantized Hall effect and the localization effect in a two-dimensional system in a strong magnetic field. A brief description will be given of the fractional quantum Hall effect and the ground state of the two-dimensional system in a strong magnetic field.

In Sec. 2 a brief introduction is given to the characteristics of the transport in two-dimensional systems and to the nature of the quantum Hall effect. Examples of experimental results are presented and the application of the quantum Hall effect will be discussed in Sec. 3. There have been several attempts to relate the quantization of the Hall conductivity to a topological invariant called the winding number. In Sec. 4 we shall briefly review one such attempt. Section 5 deals with results of extensive numerical studies on the Anderson localization and its relation to the quantization of the Hall conductivity. The fractional quantum Hall effect, in which the Hall conductivity is quantized into fractions of $-e^2/h$, is briefly discussed in Sec. 6.

2. ELECTRONS IN TWO-DIMENSIONAL SYSTEMS

The Hamiltonian of a two-dimensional system in a magnetic field of strength B is given by

$$H = -\frac{\hbar^2}{2m}\frac{\partial^2}{\partial x^2}+\frac{\hbar^2}{2m}\left(\frac{\partial}{i\partial y}+\frac{eB}{\hbar}x\right)^2 , \tag{2.1}$$

where we have chosen the Landau gauge, i.e., $\mathbf{A}=(0,Bx)$. Since the Hamiltonian does not contain the y coordinate explicitly, it has a plane wave $\exp(ik_y y)$ as a solution and reduces to that of a harmonic oscillator with its center $X=-\ell^2 k_y$, where ℓ is the cyclotron radius of the lowest Landau level defined by $\ell^2=\hbar/eB$. The energy eigenvalues are given by $E_N=(N+1/2)\hbar\omega_c$, with $\omega_c=eB/m$. Let us consider a system having a rectangular form with the size L_x and L_y and impose periodic boundary conditions in the y direction. Then, the center coordinate X takes the discrete value $j\Delta X$ with $j=1,2,..N_X$, where $\Delta X=2\pi\ell^2/L_y$ and $N_X=L_xL_y/2\pi\ell^2$.

The coordinates (x,y) of an electron can be expressed as a sum of relative (ξ,η) and center (X,Y) coordinates by $x=\xi+X$ and $y=\eta+Y$ with $\xi=(\ell^2/\hbar)(p_y+eA_y)$ and $\eta=-(\ell^2/\hbar)(p_x+eA_x)$. In the presence of potential $V(\mathbf{r})$, the center coordinates are no longer a constant of motion and follow the equation of motion:

$$\dot{X}=\frac{1}{i\hbar}[X,V]=\frac{l^2}{\hbar}\frac{\partial V}{\partial y},\quad \dot{Y}=-\frac{l^2}{\hbar}\frac{\partial V}{\partial x} . \tag{2.2}$$

If $V(\mathbf{r})$ is replaced by $eE_y y$ corresponding to a constant electric field E_y, we get the well-known result that the center of cyclotron orbit moves in the x direction with a constant velocity $v_x=E_y/B$. Another way of stating this fact is that the electric field E_y disappears in a frame moving in the negative x direction with the speed E_y/B. Therefore, the diagonal conductivity σ_{xx} vanishes and the Hall conductivity σ_{xy} becomes $-ne/B$ with n the electron concentration in a unit area.

In the presence of scatterers such as impurities, the degeneracy ($1/2\pi\ell^2$ per unit area) with respect to the center coordinates is lifted, and the density of states, which is given by a series of δ-functions in the absence of scatterers, has a nonzero width. Furthermore, the diagonal conductivity σ_{xx} becomes nonzero and the Hall conductivity is modified by a frictional force due to scattering.

The peak value of σ_{xx} is known to be given by $(N+1/2)e^2/\pi^2\hbar$ in case of short-range scatterers. This can be understood in terms of a simple diffusion picture [4,5]. In strong magnetic fields the conduction takes place because of jumps of the center coordinates as is schematically illustrated in Fig. 1. The distance of each jump is of the order of the radius of the cyclotron orbit and the diffusion constant D^* of the center of the orbit is given by $D^*\sim(2N+1)\ell^2/\tau$ where τ is the lifetime of the orbit. Let Γ be the width of the Landau level. Then, the uncertainty relation $\Gamma\tau\sim\hbar$ holds and the density of states is of the order of $D(E_F)\sim(2\pi\ell^2\Gamma)^{-1}$. The Einstein relation gives

$$\sigma_{xx} = e^2D^*D(E_F) \sim \frac{e^2}{\pi^2\hbar}(N+\frac{1}{2}) . \tag{2.3}$$

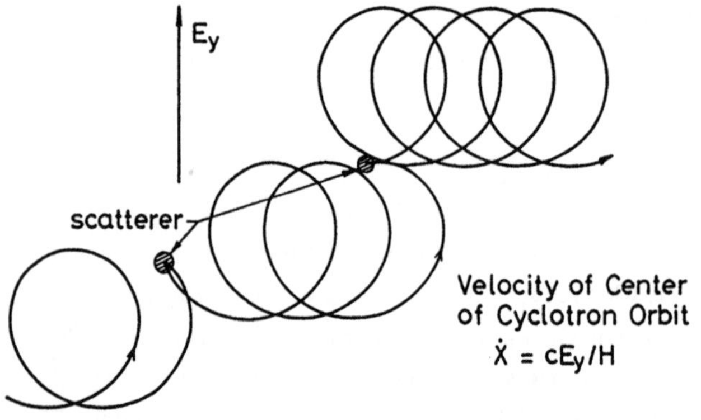

Fig. 1 Motion of an electron in a strong magnetic field

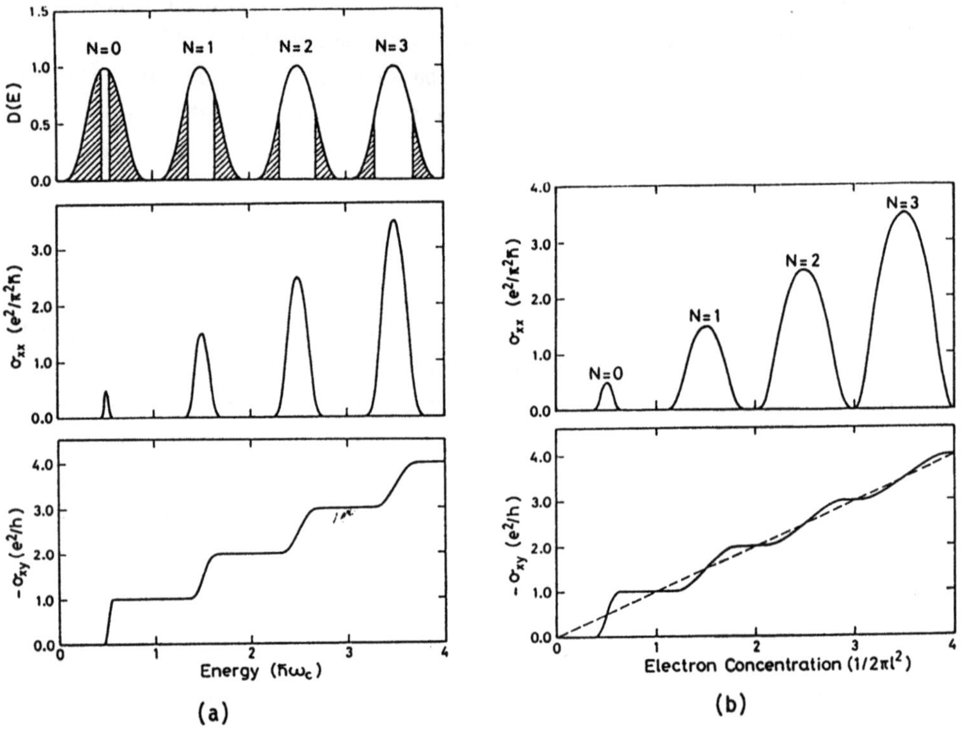

Fig. 2 The density of states D(E), the diagonal σ_{xx} and Hall σ_{xy} conductivity in a two-dimensional system in a strong magnetic field as a function of energy (a) and electron concentration (b). The dashed regions correspond to localized states, where σ_{xx} vanishes and σ_{xy} is quantized into integer multiples of $-e^2/h$

Therefore the peak value depends only on the Landau level index N and is independent of relevant quantities such as the level width and the strength of the magnetic field. This fact has been confirmed experimentally.

In addition to the broadening of Landau levels, some of the states, especially those which lie in the tail region of each Landau level, are likely to be localized spatially, i.e., the wave function is not extended throughout the system but confined within some finite region around a certain localization center. Such a localization phenomenon due to random potential fluctuations is known to be the Anderson localization and has been a subject of considerable study. Localized states do not contribute to the current conduction and the diagonal conductivity vanishes when the Fermi level lies in localized states. The Hall conductivity, on the other hand, vanishes when the Fermi level lies in localized states below the ground Landau level. When it crosses extended states and reaches upper localized states of the Landau level, σ_{xy} reaches and remains $-e^2/h$ as long as the Fermi level lies in the localized states. Such a quantization of the Hall conductivity into integer multiples of $-e^2/h$ is called the quantum Hall effect and has attracted much attention recently both experimentally and theoretically. Figure 2 schematically shows the density of states, σ_{xx}, σ_{xy}, and the quantization into $-e^2/h$ in the two-dimensional system in a strong magnetic field.

3. TWO-DIMENSIONAL SYSTEMS

One of the most typical two-dimensional systems is the inversion layer produced on semiconductor surfaces. An inversion layer is formed, when a sufficiently strong electric field is applied across the interface of an insulator and a p-type semiconductor and when the conduction band edge is bent below the Fermi level in the bulk. The motion in the z-direction, the direction perpendicular to the surface, is quantized and the energy levels of the electrons are grouped in electric subbands, each of which corresponds to a quantized level for motion in the z direction, with a continuum for motion in the xy plane parallel to the surface. It is possible only for the lowest electric subband to be occupied at sufficiently low temperatures, and we can regard the inversion layer essentially as a two-dimensional system. Figure 3 schematically shows the structure of an inversion layer. One of the important features of the inversion layer is that the electron density can be freely controlled by changing the bias voltage (called the gate voltage), since it is proportional to the electron concentration in the inversion layer. The quantum Hall effect was observed for the first time in such a system.

Similar two-dimensional systems can be realized at a heterostructure of different semiconductors like GaAs and AlGaAs, as is shown in Fig. 4. In this system the AlGaAs layer plays the role of a barrier potential. The heterostructure consisting of GaAs and AlGaAs has a supreme feature that it has extremely good interface quality. In contrast to inversion layers on Si surfaces, where the barrier consists of amorphous SiO_2, the electron mobility is extremely high (as high as 10^6 cm^2/Vs in contrast to at most 10^4 cm^2/Vs in Si inversion layers). In this system, effects of potential randomness play a minor role in comparison with those of mutual electron-electron interactions, and the fractional quantum Hall effect has been observed. See Sec. 6 for more details on the fractional quantum Hall effect.

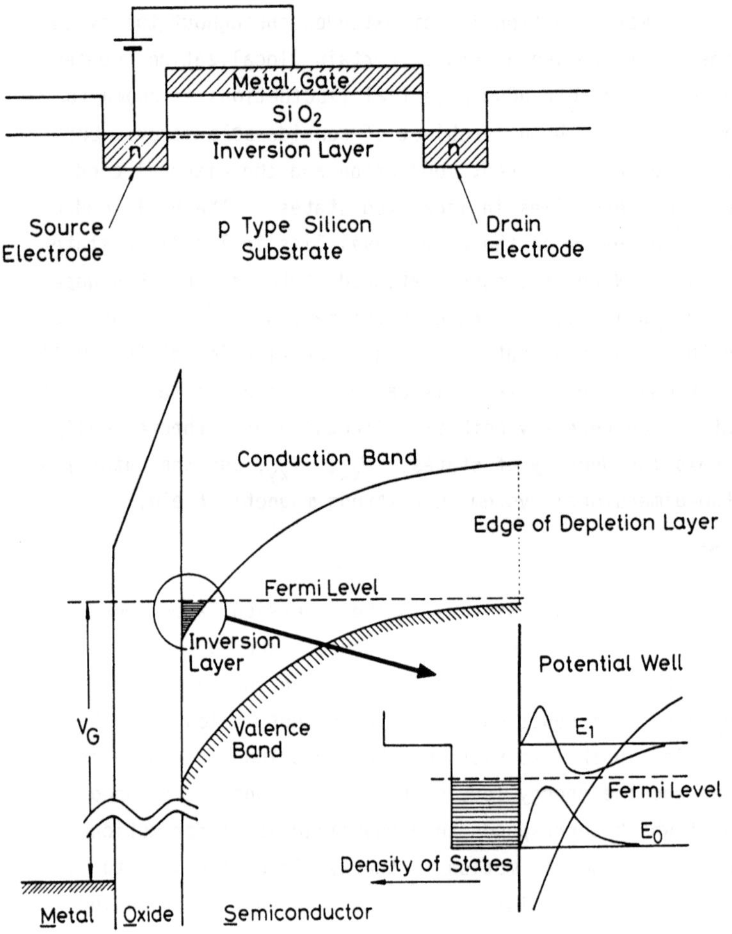

Fig. 3 Metal-oxide-semiconductor (MOS) structure, an inversion layer and energy levels of two-dimensional electrons

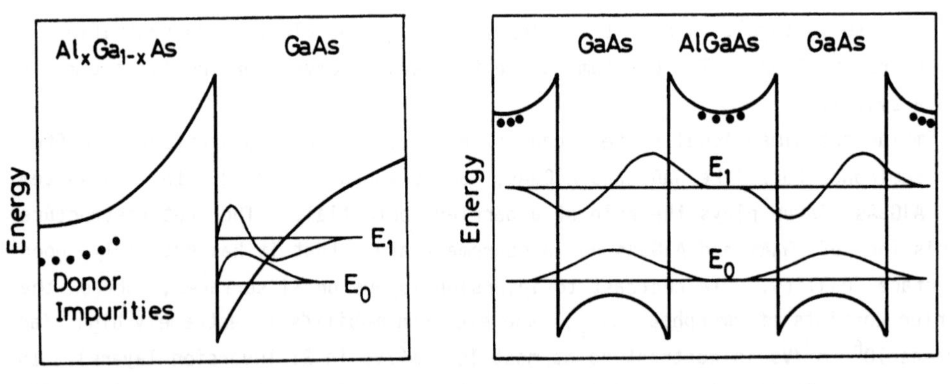

Fig. 4 Schematic illustration of GaAs/AlGaAs heterostructure and superlattice

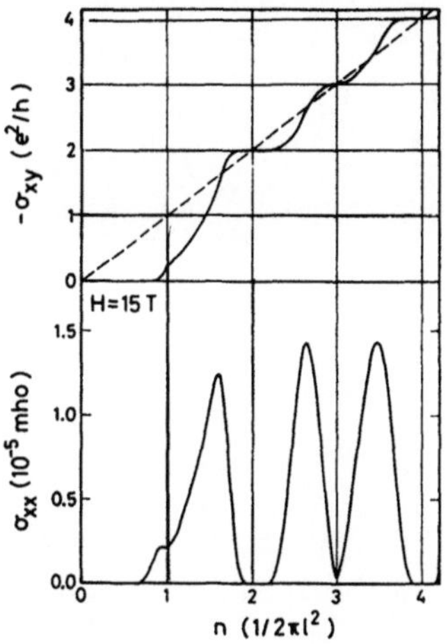

Fig. 5 An example of the diagonal conductivity (lower panel) and the Hall conductivity (upper panel) observed in a Si inversion layer by KAWAJI and WAKABAYASHI [6]. The dashed line represents ne/B with n the electron concentration. Only the lowest N=0 Landau levels (split into four by spin-Zeeman and valley splittings) are shown

Figure 5 gives an example of the beautiful experimental results observed by KAWAJI and WAKABAYASHI in the inversion layer on the Si (100) surface [6]. The lowest (N=0) Landau level splits into four by the removal of spin and valley degeneracy. The conduction band of Si has six equivalent minima in the [001] and equivalent directions. A part of this degeneracy is lifted due to the strong surface electric field, but the ground electric subband is still doubly degenerate in the effective-mass approximation. Actually, this degeneracy is also lifted slightly.

VON KLITZING, PEPPER, and DORDA have measured the Hall resistance instead of the Hall conductivity and demonstrated that the Hall resistance is quantized into integer multiples of h/e^2 [2]. This is a direct consequence of the fact that the Hall conductivity is quantized into an integer times $-e^2/h$ and the diagonal conductivity vanishes, since the resistivity tensor is related to the conductivity tensor through

$$\rho_{xx} = \frac{\sigma_{xx}}{\sigma_{xx}^2+\sigma_{xy}^2}, \quad \rho_{xy} = -\frac{\sigma_{xy}}{\sigma_{xx}^2+\sigma_{xy}^2} . \tag{3.1}$$

The most important feature of the quantum Hall effect is that the quantized Hall resistance is exactly an integer multiple of h/e^2 irrespective of sample geometries whenever the diagonal resistance vanishes. Note that the resistance (conductance) and resistivity (conductivity) have the same dimensionality in two dimensions, while they are different by an extra length scale in three dimensions.

Since h/e^2 is a universal quantity having the dimension of resistance, the quantized Hall resistance can be used as its standard. Further, by measuring h/e^2 to high accuracy and combining it with the accurately measurable light velocity c, one can determine the fine structure constant $e^2/4\pi\varepsilon_0\hbar c$ ($\sim$1/137). Such a high-precision measurement of the quantized Hall resistance has been performed by several groups in different countries. Figure 6 shows the fine structure constant determined by the high-precision measurement of the quantum

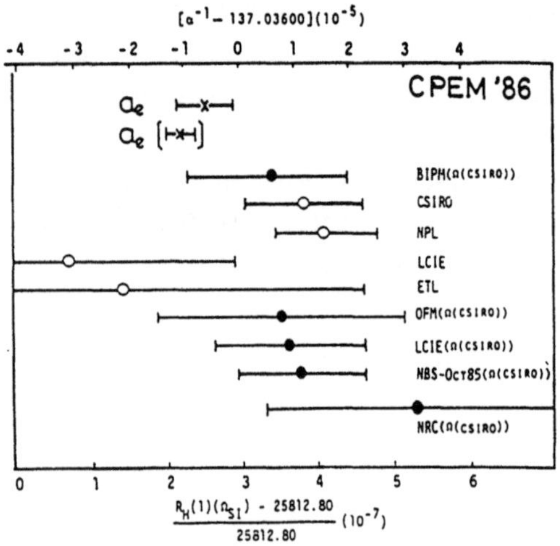

Fig. 6 The inverse of the fine structure constant, $4\pi\varepsilon_0\hbar c/e^2$ ($\sim$137), determined by high-precision measurement of quantized Hall resistance and that determined through anomalous magnetic moment of a free electron combined with elaborate calculations based on the quantum electrodynamics by KINOSHITA. Closed circles represent results obtained based on 1Ω maintained by CSIRO (Australia). Data are from NBS (U.S.A.), BIPM (France), NPL (England), ETL (Japan), OFM (Switzerland), NRC (Canada), and CSIRO. There seem to be slight discrepancies between the fine structure constant determined by the quantized Hall resistance and that determined by the anomalous magnetic moment. However, inaccuracies of existing resistance standards still make it impossible to draw any definite conclusions. After Ref. 7

Hall effect [7]. The figure shows that the accuracy of the measurement of the quantized resistance is not enough to determine whether the quantum electrodynamics (QED) is valid or not. The main origin of such inaccuracies is uncertainty in the resistance standards maintained in various countries. Note also a possibility of systematic errors in resistance standards. In the conventional method, the resistance standard is defined through capacitance of a cross-capacitor, whose capacitance can be calculated quite accurately.

4. TOPOLOGICAL INVARIANTS

The quantization of the Hall conductivity was first noted by ANDO, MATSUMOTO, and UEMURA in 1975 [8], although effects of localization were not discussed explicitly [3]. Later in 1982, LAUGHLIN [9] provided a convincing argument, which shows that the Hall conductivity is quantized into an integer multiple of $-e^2/h$ whenever the Fermi level lies in localized states. There appeared various modified versions of Laughlin's argument. A rigorous mathematical proof based on the Kubo formula appeared only recently. Here, a brief review will be given on one of such attempts in which the Hall conductivity in the torus geometry averaged over magnetic fluxes passing through torus is always given by an integer times $-e^2/h$ [10-12].

We consider a two-dimensional system with a finite size LxL in a magnetic field B. The Hall conductivity σ_{xy} is written in terms of the velocity operators v_x and v_y as

$$\sigma_{xy} = \frac{\hbar e^2}{i\pi L^2}\int dEf(E)\mathrm{Tr}\,[v_x(\frac{\partial}{\partial E}\mathrm{Re}G(E+i0))v_y\mathrm{Im}G(E+i0)-(x\leftrightarrow y)]\,, \quad (4.1)$$

where $G(E)=(E-H)^{-1}$ is Green's function or a resolvent, E is energy, and H is the Hamiltonian. Now, we introduce a vector potential $\mathbf{A}^0=(A_x{}^0,A_y{}^0)$ independent of position and time. We can easily show that the above can be rewritten as

$$\sigma_{xy}=-\frac{\hbar}{4\pi L^2}\int_{E_f-i\infty}^{E_f+i\infty}dE\ \mathrm{Tr}\,[G\frac{\partial G^{-1}}{\partial A_x^0}G\frac{\partial G^{-1}}{\partial A_y^0}G\frac{\partial G^{-1}}{\partial E}-(x\leftrightarrow y)]\,,\quad v_x=\frac{\partial H}{e\partial A_x^0}\,. \quad (4.2)$$

The energy and conductivities are periodic in $A_x{}^0$ and $A_y{}^0$ with a period h/eL. When the system size is sufficiently large, the physical quantities do not depend on the vector potential $\mathbf{A}^0$ and the observed conductivity is equal to the average over the vector potential $\mathbf{A}^0$. We have

$$\sigma_{xy}=-\frac{\hbar}{4\pi L^2}(\frac{eL}{\hbar})^2\int dE\int dA_x^0\int dA_y^0\mathrm{Tr}\,[G\frac{\partial G^{-1}}{\partial A_x^0}G\frac{\partial G^{-1}}{\partial A_y^0}G\frac{\partial G^{-1}}{\partial E}-(x\leftrightarrow y)]\,. \quad (4.3)$$

Now, the quantity

$$-\frac{1}{8\pi^2}\int dE\int dA_x^0\int dA_y^0\mathrm{Tr}\,[G\frac{\partial G^{-1}}{\partial A_x^0}G\frac{\partial G^{-1}}{\partial A_y^0}G\frac{\partial G^{-1}}{\partial E}-(x\leftrightarrow y)] \quad (4.4)$$

is known to be a topological invariant called "the winding number" which appears in non-Abelian gauge theory and is always an integer.

The vector potential $\mathbf{A}^0$ may be thought of as two magnetic fluxes, $(\Phi_x,\Phi_y) = (A_x{}^0L,A_y{}^0L)$, which penetrate respectively inside and through the opening of a torus when we impose periodic boundary conditions in both x and y directions for the system of size L. The physical system assumes its original state every time $A_x{}^0$ or $A_y{}^0$ increases by Φ_0/L, with $\Phi_0=h/e$ being the magnetic-flux quantum.

We have shown that the Hall conductivity averaged over the flux is always an integer multiple of $-e^2/h$. This does not necessarily mean, however, that the observed Hall conductivity is quantized. Actually, the integer depends on the Fermi level and varies whenever the Fermi level crosses extended states distributed densely in macroscopic systems. Therefore, the observed Hall conductivity is its average and needs not be quantized into the integer times $-e^2/h$. When the Fermi level lies in localized states, the localized states in the vicinity of the Fermi level do not contribute to the Hall conductivity, and consequently the averaged Hall conductivity becomes also the integer multiple of $-e^2/h$. The quantization of the Hall conductivity into the integer multiple of $-e^2/h$ is, therefore, a direct consequence of the localization and the mathematical theorem that the winding number is an integer.

Note here that nothing can be said about the quantized value of the Hall conductivity as Laughlin's original argument [9] could not predict the actual value of the integers. Note also that the above theoretical explanation does not explicitly take into account the two-dimensionality of the system. In three-dimensional systems similar arguments show that the Hall conductance is quantized into integer multiple of $-e^2/h$ whenever the Fermi level lies in localized states.

5. LOCALIZATION IN STRONG MAGNETIC FIELDS

5.1 Symmetry and Dimensionality

The dimension and the symmetry of the system play important roles in the localization problem. This can most easily be understood if we consider the scaling argument. Let us consider a d-dimensional cube with a size L. By combining different cubes of the same size, we can construct a cube with size 2L. By repeating such a procedure starting from a small system, we can obtain energy levels and wave functions of a system having arbitrary size. Combination of two cubes causes mixings of energy levels. In this case couplings of levels closest to each other in energy are most important. Let V(L) be the resonance energy of such energy levels and W(L) their energy difference. We can safely assume W(L) to be the average energy separation, i.e., $W(L) \sim 1/L^dD(E)$ with D(E) the density of states.

The ratio $g(L)=V(L)/W(L)$ is called the Thouless number. If states are localized, g(L) decreases with the increase of the system size L reflecting how the wave function decays with the distance from the localization center. For extended states g(L) is proportional to the conductance of the cube. That is, $(e^2/\hbar)g(L)$ is the conductance and approaches $L^{d-2}\sigma$ in the limit of infinitely large L, where σ is the conductivity.

It can readily be understood that the localization depends strongly on the dimension and the symmetry of the system. When the wave function can be chosen as real, i.e., the Hamiltonian is a real symmetric matrix, the resonance integral describing couplings between cubes is real, i.e., the "dimension" of the interaction is $\beta=1$. This corresponds to the system in the absence of a magnetic field. When a magnetic field is applied, the time-reversal symmetry is broken and the wave function becomes complex or the Hamiltonian a complex Hermitian matrix. Then we have $\beta=2$. If we consider electron spins, we have another symmetry. In the presence of spin-orbit interactions, interactions are now allowed between degenerate levels and $\beta=4$. We call $\beta=1$ orthogonal ensemble, $\beta=2$ unitary, and $\beta=4$ symplectic. It is clear that states are less easily localized for larger β.

ABRAHAMS et al. [13] introduced the single-parameter scaling theory in 1979. In their theory with g(L) being the scaling variable, states are always localized in two dimensions. Fortunately, the theory of AOKI and ANDO [14] shows that the single-parameter scaling theory is invalid and there should exist some delocalized states in each Landau level at least in the strong-field limit, where mixings among Landau levels can be neglected.

5.2 Numerical Study

One of the most direct approaches to study the localization problem is numerical computer simulations. Although the system size considered is limited, an appropriate criterion of localization enables the extrapolation to infinitely large systems. One of the most powerful methods is to use the Thouless number g(L) and it has been employed in extensive investigations for the determination of the energy dependence of the inverse localization length $\alpha(E)$ [15].

We consider a system with a finite size LxL. Scatterers are distributed at random. Periodic boundary conditions are used for both x and y directions. The coupling V(L) can be estimated by the energy shift ΔE due to the change in boundary conditions, i.e., by replacing periodic boundary conditions in the y direction by antiperiodic conditions. The coupling V(L) is the geometric mean of ΔE of a given energy interval and over different samples.

Another powerful approach is the finite-size scaling method. Consider a two-dimensional system having a form of long strips, $L_x x L_y$ ($L_x >> L_y$). The wave function is always localized exponentially in the x direction, since the system becomes essentially one-dimensional as long as $L_x >> L_y$. Let $\alpha(L_y)$ be the inverse localization length in the y direction for a given L_y. We can determine $\alpha(E)$ by detailed study of L_y dependence. This method has also been applied to the present case [16].

Figure 7 shows the resulting inverse localization length as a function of energy for systems containing short-range scatterers of high concentration. The results obtained by the two independent methods, i.e., the Thouless number and the finite-size scaling, agree quite well. The figure contains also the results for the first excited Landau level N=1, which have been obtained in similar procedures.

It is widely accepted that states are all localized in two dimensions in the absence of a magnetic field. When a weak magnetic field is applied to the system, the magnetic field tends to increase the extent of localized wave functions. This effect gives rise to the negative magnetoresistance widely observed in various quasi-two-dimensional systems. It is believed, however, that the magnetic field does not destroy the localization as long as it is sufficiently weak. Therefore, an important question arises: What happens when mixing between different Landau levels is appreciable with decreasing magnetic field? The Thouless-number method can be extended to study such effects of level mixing.

Figure 8 shows the histogram of the density of states for the lowest Landau level [15]. The three lowest Landau levels N=0,1,2 have been included in the calculation. For $\hbar\omega_c/\Gamma=2.5$ (ω_c is the cyclotron frequency), the density of states

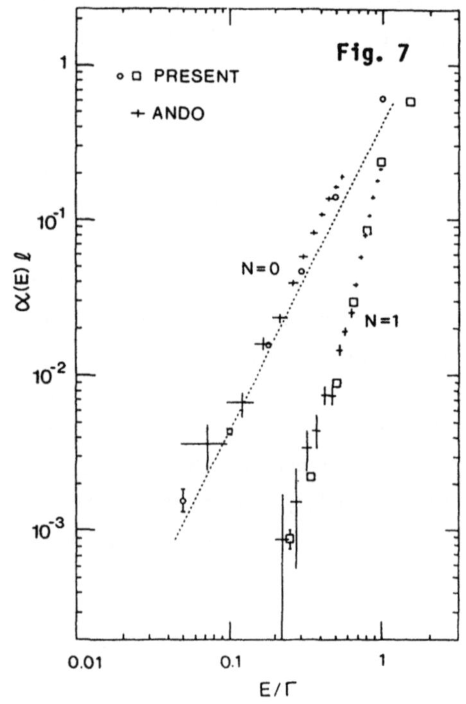

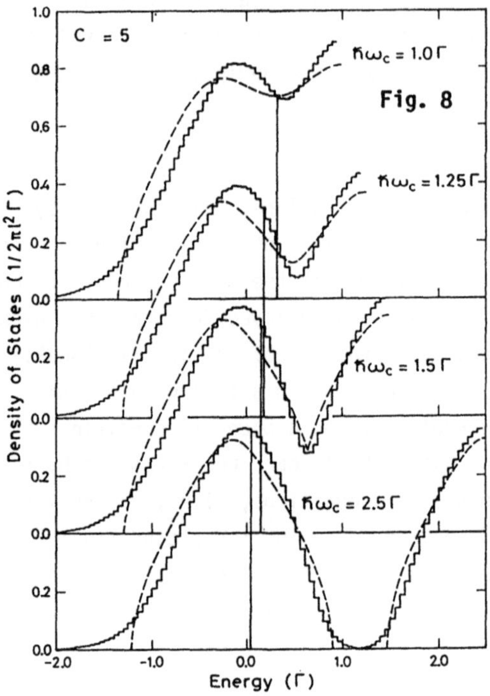

Fig. 7 Log-log plot of the inverse localization length $\alpha(E)$ in units of the inverse of the cyclotron radius of the lowest Landau level for N=0 and 1 versus energy measured from the center of each Landau level and normalized by the broadening Γ calculated in the self-consistent Born approximation. Scatterers with short-range potentials are distributed densely in the system. The crossed error bars are the result of the Thouless number study [15] and the squares that of the finite-size scaling study [16]

Fig. 8 Histograms of the density of states when mixing among different Landau levels is taken into account [15]. The dashed lines represent the results calculated in the single-site approximation. The energy is normalized by the broadening Γ calculated in the single-site approximation and its origin is chosen at the position of the lowest Landau level. The vertical straight lines represent the energy of delocalized states, i.e., the energy where $\alpha(E)$ vanishes

of each Landau level is separated from each other in the single-site approximation. On the other hand, the numerical density of states has a small tail and does not vanish in the gap region. With decreasing magnetic field, the energy corresponding to the peak of the density of states is shifted to the lower energy side due to the quantum-mechanical level-repulsion effect. In contrast to this, the energies of extended states, denoted by vertical straight lines, are shifted to the higher energy side.

This result suggests the interesting possibility that there always exist some extended states in the two-dimensional system in magnetic fields. Those extended states move up in energy with decreasing magnetic fields and disappear only in the limit of vanishing magnetic field.

5.3 Scaling Relation between Conductivities

The system-size dependence of the Hall conductivity can also be studied numerically [17,18]. It has been shown that with increasing L, the Hall conductivity approaches a step function. Figure 9 shows $\sigma_{xx}(E,L)$, proportional to the Thouless number $g(E,L)$, versus $\sigma_{xy}(E,L)$ for a system with short-range scatterers. The diagonal and off-diagonal conductivities for different E's and system-size L's are plotted in the same diagram. The figure clearly demonstrates that σ_{xx} and σ_{xy} are mutually correlated and cannot be considered independent in the present system. However, this scaling relation between σ_{xx} and σ_{xy} is dependent on N. Furthermore, mixing of different Landau levels, which is important in weaker magnetic fields, has been shown to drastically modify the scaling relation, i.e., the σ_{xx} versus σ_{xy} curve does depend on the ratio $\Gamma/\hbar\omega_c$, where Γ is the level broadening and $\hbar\omega_c$ is the energy separation between adjacent Landau levels.

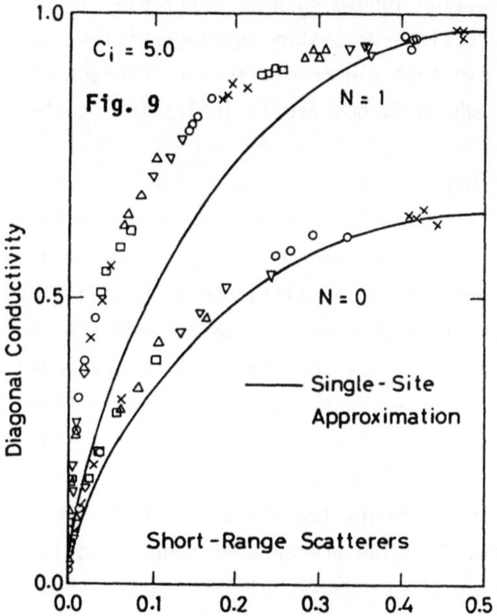

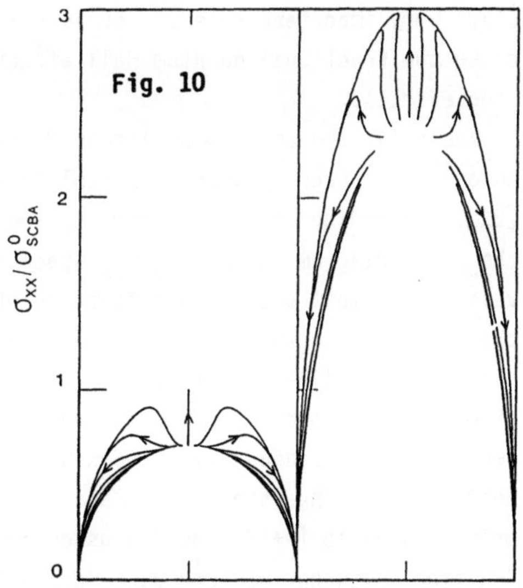

Fig. 9 Flow diagrams for σ_{xy} and σ_{xx} for the lowest two Landau level [18]. The Thouless number, proportional to σ_{xx}, and the Hall conductivity of different energies and four different system sizes L's are plotted

Fig. 10 Calculated flow diagram for $\sigma_{xy}(T)$ and $\sigma_{xx}(T)$ for varying temperature in a model system [19]

This scaling relation between conductivities may possibly be observed experimentally. One way is to study the temperature "flow diagram". This may be possible since inelastic scattering present at nonzero temperatures introduces an effective system size and by varying temperature we can study the system-size dependence of the conductivities. Figure 10 gives an example of calculated flow lines [19]. The envelope of flow lines for different Fermi levels becomes the scaling function of the conductivities.

KAWAJI [20] has shown that temperature flow lines observed in Si inversion layers in the temperature range between 1.5 and 4.2 K are consistent with the curve given in Fig. 10. WEI et al. reported that the temperature flow lines in two-dimensional systems at InGaAs/InP heterostructures are similar to those shown in Fig. 10 in a similar temperature range but deviate drastically at extremely low temperatures (∿50 mK) [21]. The equivalence of the scaling function between σ_{xx} and σ_{xy} and the temperature flow line mentioned above is based on a drastic and simple-minded assumption that the only effect of electron-electron interactions is to introduce a cut-off in the effective system size because of inelastic scatterings. It is quite likely that this assumption is invalid although it may be valid in some narrow temperature range. Especially when the temperature is extremely low and consequently the localization length becomes extremely large, other effects like electron-electron interactions or system inhomogeneities can play very important roles. It is now known that electron-electron interactions cause the fractional quantum Hall effect, which is not merely inelastic scattering effects.

Recently, the unitary nonlinear σ model with σ_{xy} as a θ term in the Yang-Mills theory has been proposed to explain the localization and quantum Hall effect [22]. According to this theory σ_{xx} and σ_{xy} are independent scaling variables and the flow diagram in the σ_{xy}-σ_{xx} space determines the localization in Landau levels. The most significant features of the flow diagram are the periodicity in the σ_{xy} axis, the presence of unstable fixed points corresponding to extended states on the line $\sigma_{xy}=-(N+1/2)e^2/h$, and its independence of the magnetic field strength. It predicts that the peak value of the diagonal conductivity is independent of the Landau level index N, which is in clear contradiction with experiments in Si inversion layers and intuitive arguments leading to the peak value proportional to (N+1/2) as discussed in Sec. 2. The presence of single scaling function between σ_{xx} and σ_{xy}, dependent on the Landau-level index N in strong magnetic fields, and also strong effects of level mixing on the scaling function clearly demonstrate the invalidity of the flow diagram proposed in the nonlinear σ model.

6. GROUND STATE

So far we have not explicitly considered effects of mutual interactions between electrons. In this section we shall discuss the ground state of the two-dimensional system in strong magnetic fields and its relation to the fractional quantum Hall effect.

The fractional quantum Hall effect was observed for the first time by TSUI, STORMER, and GOSSARD in a two-dimensional system at GaAs/AlGaAs heterostructures [23,24]. It was later observed also in inversion layers on Si surfaces. In the fractional quantum Hall effect, the Hall conductivity is quantized into fractions of $-e^2/h$, when the Fermi level lies in the lowest Landau level N=0. When the Hall conductivity is quantized, the diagonal conductivity σ_{xx} shows an activated temperature dependence, i.e., $\sigma_{xx} \propto \exp(-\Delta E/k_B T)$, where ΔE is an activation energy, T is the temperature, and k_B is the Boltzmann constant. The observed fractions are 1/3, 2/3, 2/7, 2/5, ... It is very clear that this fractional quantum Hall effect cannot be explained by the localization effect as the ordinary (integral) quantum Hall effect.

It is now widely believed that the ground state in the strong field limit is described by approximate wave functions proposed by LAUGHLIN [25]. In the following, we shall use the symmetric gauge, i.e., $\mathbf{A}=(-By/2,Bx/2)$, and choose as a basis wave functions in which angular momentum around the origin is diagonalized. The basis wave function is written as

$$\varphi_m(x,y)=(2\pi 2^m m!)^{-1/2} z^m \exp\left(-\frac{|z|^2}{4}\right), \tag{6.1}$$

where $z=(x+iy)/\ell$ and the angular momentum is m with m=0,1,..,M-1. We consider a circular system with area $S=2\pi\ell^2(M-1)$.

When N electrons are confined within this system, the wave function is given by a linear combination of Slater determinants determined by (6.1). In general, we have

$$\Psi(z_1,\ldots,z_N) \propto \prod_{i<j}(z_i-z_j) f(z_1,\ldots,z_N) \exp\left(-\sum_{i=1}^{N}\frac{|z_i|^2}{4}\right), \tag{6.2}$$

where $f(z_1,..,z_N)$ is a symmetric polynomial. Electrons are mutually separated far apart in the ground state because of the Coulomb repulsive force. Now, let us consider the following wave function:

$$\Psi_{1/p}(z_1,\ldots,z_N) \propto \prod_{i<j}(z_i-z_j)^p \exp\left(-\sum_{i=1}^{N}\frac{|z_i|^2}{4}\right), \tag{6.3}$$

where the power p should be an odd integer because of the antisymmetry of the wave function. In the state described by (6.3), relative angular momentum of all electron pairs is p and the pair distribution function $\rho(r)$ approaches zero in proportion to r^{2p} near the origin. That is, the state described by this wave function, called the 1/p Laughlin state, is considered to posses a low energy and be close to the ground state.

Such wave functions do not correspond to the ground state in general but only for a special value of the level occupation ratio ν defined by $\nu=2\pi\ell^2 n$, where n is the electron concentration. The maximum momentum M-1 or the number of basis states is obtained from the maximum power of z_1 as $M=(N-1)p+1$. That is, the level occupation should satisfy the condition $\nu=N/M\leqslant 1/p$ for sufficiently large M and N. This means that such a wave function cannot be realized when $\nu>1/p$. Further, this state is not a ground state for $\nu<1/p$, since in this case we can construct states having a pair of electrons with a mutual angular momentum larger than p. Therefore, we can easily surmise that the Laughlin state becomes a ground state only at $\nu=1/p$.

Some of the properties of the Laughlin state can be seen from

$$|\Psi_{1/p}(z_1,\dots,z_N)|^2 \propto \exp\Big[2p\sum_{i>j}\ln|z_i-z_j| - \sum_{i=1}^{N}\frac{|z_i|^2}{2}\Big] . \tag{6.4}$$

The right-hand side is the same as the Boltzmann factor of a classical two-dimensional Coulomb gas (a system of line charges interacting via a potential depending on the distance logarithmically). The first term in the exponential function represents the interaction between line charges and the second is the interactions with a uniform background charge. It is known that the states correspond to a liquid phase with no long-range order except in the case that $p\gg 1$.

Low-energy excited states can also be constructed from the above wave function. For example, an excited state corresponding to a "quasi-hole" at z_0 is given as

$$\Psi_{1/p}(z_1,\dots,z_N;z_0) \propto \prod_{i<j} (z_i-z_j)^p(z_i-z_0) \exp\Big[-\sum_{i=1}^{N}\frac{|z_i|^2}{4}\Big] . \tag{6.5}$$

It is easily seen that the total electron-density deficiency in the vicinity of z_0 is smaller than that corresponding to a single electron. As a matter of fact, the pair distribution function in the vicinity of the origin vanishes as r^{2p} corresponding to the Coulomb hole while the density near z_0 is only proportional to r^2. A simple calculation shows that the total charge deficiency around z_0 is only $+e/p$, meaning that a quasi-hole has a fractional charge $+e/p$. Similarly, we can construct an excited state having a quasi-electron with a fractional charge $-e/p$. Note that the remaining charge $(p-1)e/p$ is distributed around the system edge. Since the Laughlin wave function is "rigid", we can easily surmise that it has a nonzero energy gap between the excited states with fractionally charged quasi-electrons or quasi-holes. There have been many calculations of the energy gap and at present the gap is believed to be about $0.1\times e^2/4\pi\varepsilon_0\ell$ for the lowest Landau level.

It is now almost clear that the system can have a special ground state at $\nu=1/p$ with p an odd integer. We can similarly construct the ground-state wave function for $\nu=1-p^{-1}$ using the "electron-hole" symmetry around $\nu=1/2$. The wave function for $\nu=1$ is given by (6.3) with $p=1$. The state corresponding to the presence of a "hole" having angular momentum m is given by

$$\Phi_m(z_1,\ldots,z_N) \propto \int dx_{N+1}\int dy_{N+1}\ \varphi_m^*(z_{N+1})\Psi_1(z_1,\ldots,z_N,z_{N+1}) . \tag{6.6}$$

Therefore the ground state of $\nu=1-p^{-1}$ is given by the wave function describing the state in which all hole pairs possess mutual angular momentum p, i.e.,

$$\begin{aligned}\Psi_{(p-1)/p}(z_1,\ldots,z_N) \propto &\prod_{k=1}^{M-N}\int dx_{N+k}\int dy_{N+k}\prod_{N<i<j}^{M}(z_i^*-z_j^*)^p \\ &\times \exp[-\sum_{i=N+1}^{M}\frac{|z_i|^2}{4}]\ \Psi_1(z_1,\ldots,z_N,z_{N+1},\ldots,z_M) .\end{aligned} \tag{6.7}$$

We can easily see that the occupation ratio corresponding to this wave function is given by $\nu=N/M=1-p^{-1}$.

Next, let us consider the following wave function,

$$\Psi_\nu(z_1,\ldots,z_N) \propto \prod_{i<j}(z_i-z_j)^q\ \Psi_{(p-1)/p}(z_1,\ldots,z_N) , \tag{6.8}$$

with q being an even integer. This wave function is again expected to describe a ground state, since electrons are mutually separated. The corresponding occupation ratio is obtained through the maximum power $q(N-1)+M-1$ of z_1 as

$$\nu = \frac{N}{q(N-1)+M} = [q+\frac{p}{p-1}]^{-1}, \tag{6.9}$$

which gives, for example $\nu=2/7$ for $p=3$ and $q=2$.

Equation (6.7) for $p=3$ can be rewritten as follows:

$$\Psi_{2/3}(z_1,\ldots,z_N) \propto g_{1/3}(z_1,\ldots,z_N)\prod_{i<j}(z_i-z_j)\ \exp[-\sum_{i=1}^{N}\frac{|z_i|^2}{4}] . \tag{6.10}$$

The symmetric polynomial $g_{1/3}$ is regarded as an "operator" creating a 1/3 Laughlin state of holes. Similarly (6.8) for $q=2$ can be written as

$$\Psi_{2/7}(z_1,\ldots,z_N) \propto g_{1/3}(z_1,\ldots,z_N)\prod_{i<j}(z_i-z_j)^3\ \exp[-\sum_{i=1}^{N}\frac{|z_i|^2}{4}] , \tag{6.11}$$

which means that the $\nu=2/7$ wave function describes the state in which quasi-holes in the 1/3 Laughlin state form a 1/3 Laughlin state. In the $\nu=2/5$ state, on the other hand, quasi-electrons in the 1/3 Laughlin state form a 1/3 Laughlin state, i.e.,

$$\Psi_{2/5}(z_1,\ldots,z_N) \propto \exp[-\sum_{i=1}^{N}\frac{|z_i|^2}{4}]\ g_{1/3}(\frac{\partial}{\partial z_1},\ldots)\prod_{i<j}(z_i-z_j)^3 . \tag{6.12}$$

In this way we can construct various kinds of ground-state wave functions starting from the 1/p Laughlin wave function. In particular, the hierarchy series starting from the 1/3 wave function can explain all fractions so far observed experimentally. At electron concentrations corresponding to the fractional ν's, the electron system has a special ground state having a nonzero excitation gap. Further, a small change in the electron density causes fractionally charged quasi-electrons and/or quasi-holes which are likely to be localized due to existing potential fluctuations and do not contribute to the electric current. Therefore, the fractional quantum Hall effect can be understood qualitatively. It should be noticed, however, that the quantitative understanding of experimental results especially concerning activation energies requires detailed understanding of ground state properties in the presence of both Coulomb interactions and random potential fluctuations. Note also that there have been other proposals for the ν=2/7 and 2/5 ground-state wave-functions [26].

7. SUMMARY AND CONCLUSION

A brief review has been given on the quantum Hall effect mainly from a theoretical point of view. The integral quantization of the Hall conductivity is a direct manifestation of the Anderson localization in the two-dimensional system in strong magnetic fields. The quantization can be explained theoretically by relating the Hall conductivity to a topological quantity called the winding number. The fractional quantum Hall effect is, on the other hand, explained by special ground states allowed by mutual Coulomb repulsive force among electrons. The nature of the ground state is relatively well understood, but the real understanding of the fractional quantization requires a knowledge of interplay between electron-electron interactions and random potential fluctuations.

References

1. T. Ando, A.B. Fowler and F. Stern: Rev. Mod. Phys. **54**, 436 (1982)
2. K. von Klitzing, G. Dorda and M. Pepper: Phys. Rev. Lett.**45**, 494 (1980)
3. T. Ando: Prog. Theor. Phys. Suppl. **84**, 69 (1985)
4. T. Ando and Y. Uemura: J. Phys. Soc. Jpn. **36**, 959 (1974)
5. T. Ando: J. Phys. Soc. Jpn. **36**, 1167 (1974); **37**, 622 (1975); **37**, 1233 (1975)
6. S. Kawaji and J. Wakabayashi: Physics in High Magnetic Fields, ed. by S. Chikazumi and N. Miura (Springer, Berlin, 1981), p. 284
7. S. Kawaji: Proc. 10th Int. Conf. Atomic Physics, ed. by H. Narumi (North-Holland, Amsterdam, 1987)
8. T. Ando, Y. Matsumoto and Y. Uemura: J. Phys. Soc. Jpn. **32**, 859 (1972)
9. R.B. Laughlin: Phys. Rev. B **23**, 5632 (1981)
10. Q. Niu, D.J. Thouless and Y.S. Wu: Phys. Rev. B **31**, 3372 (1985)
11. T. Ando: Statistical Physics and Condensed Matter Theory, ed. by Xie Xide (World Scientific, Singapore, 1986), p.207
12. H. Aoki and T. Ando: Phys. Rev. Lett. **57**, 3093 (1986)
13. E. Abrahams, P.W. Anderson, D.C. Licciardello and T.V. Ramakrishnan: Phys. Rev. Lett. **42**, 673 (1979)
14. H. Aoki and T. Ando: Solid State Commun. **38**, 1079 (1981)
15. T. Ando: J. Phys. Soc. Jpn. **52**, 1893 (1983); **53**, 3101 (1983); **53** 3126 (1983)
16. T. Ando and H. Aoki: J. Phys. Soc. Jpn. **54**, 2238 (1985)
17. T. Ando: Surf. Sci. **170**, 243 (1986)
18. T. Ando: J. Phys. Soc. Jpn. **55**, 3199 (1986)
19. H. Aoki and T. Ando: Surf. Sci. **170**, 249 (1986)

20. S. Kawaji: Prog. Theor. Phys. Suppl. **84**, 178 (1985)
21. H.P. Wei, A.M. Chang, D.C. Tsui, A.M.M. Pruisken and M. Razeghi: Surf. Sci. **170**, 238 (1986); H.P. Wei, D.C. Tsui and A.M.M. Pruisken: Phys. Rev. B **33**, 1488 (1985)
22. H. Levine, S.B. Libby and A.M.M. Pruisken: Phys. Rev. Lett. **51**, 1915 (1983); Nucl. Phys. B **240**, 30; 49; 71 (1984)
23. D.C. Tsui, H.L. Stormer and A.C. Gossard: Phys. Rev. Lett. **48**, 1559 (1982)
24. H.L. Stormer, A.M. Chang, D.C. Tsui, J.C. Hwang, A.C. Gossard and W. Wiegmann: Phys. Rev. Lett. 50, 1393 (1983)
25. R.B. Laughlin: Phys. Rev. Lett. **52**, 1583 (1984)
26. See, for example, B.I. Halperin: Surf. Sci. **170**, 115 (1986), and references cited therein

Diffusion of Heavy Particles in Metals

J. Kondo

Electrotechnical Laboratory, Tsukuba Research Center,
Ibarakiken 305, Japan

Protons or muons moving in the jellium of the metal electrons are treated theoretically. It is noted that the screening electron cloud cannot follow the particle adiabatically. The electrons are not fast enough, because their excitation energy can be as small as zero. When the muon or the proton tunnels from one potential well to another, the tunneling matrix is renormalized by a factor which is small compared with unity and depends on a power of the temperature. This power law reflects the non-adiabatic effect just mentioned. Measurement on the diffusion constant of the positive muon in copper as well as that of inelastic neutron scattering from protons in niobium are discussed in terms of the theory.

1. INTRODUCTION

The conduction electrons in metals are a degenerate Fermi system with the degeneracy temperature more than ten thousand degrees Kelvin. A feature characteristic of such systems is that the excitation energy can be as small as zero. This fact gives rise to an infrared divergence, which is the main topic of this talk.

First, we briefly mention how to describe the metal electrons. The simplest model for them is the free particle moving in a potential of macroscopic scale, which is constant within the metal. The electron wave function is then represented by a plane wave, which is specified by its wave vector k and whose energy is expressed by

$$\varepsilon_{\mathbf{k}} = \hbar^2 k^2/2m. \tag{1}$$

The energy levels are distributed quasi-continuously, and each lower level is occupied by two electrons with opposite spins. For the electron density of metals, the highest occupied level (the Fermi level) is more than 1eV above the bottom. At finite temperatures, there are excitations of electrons. The typical excitation energy is about $k_B T$, where T is the temperature and k_B is the Boltzmann constant. Thus, below room temperature, the metal electrons are highly degenerate. A feature characteristic of such systems is that the excitation energy can be as small as zero. This is in sharp contrast to insulators, where there is a gap for the excitation energy. We shall see in this talk what effect such vanishing excitation energy causes.

2. IMPURITY POTENTIAL IN METALS

The simplest way to see the effect caused by the vanishing excitation energy is to consider an impurity introduced into a metal. The impurity will introduce a potential energy V(r) for the electrons. The total wave function for the electrons will then be expressed by

$$\Psi = \frac{1}{N}\Big[\Phi_o - \sum_{\substack{k<k_F \\ k'>k_F}} \frac{V_{\mathbf{kk'}}}{\varepsilon_{\mathbf{k'}} - \varepsilon_{\mathbf{k}}} \Phi_{\mathbf{k}\to\mathbf{k'}} + \cdots\Big] , \qquad (2)$$

where Φ_o is the ground state (the Fermi sphere state), $\Phi_{\mathbf{k}\to\mathbf{k'}}$ is obtained from it by exciting an electron **k** to an empty state **k'**. (k_F is the Fermi wave number.) $V_{\mathbf{kk'}}$ is the matrix element of the potential, and N is the normalization constant expressed by

$$N = 1 + \sum_{\substack{k<k_F \\ k'>k_F}} \frac{|V_{\mathbf{kk'}}|^2}{(\varepsilon_{\mathbf{k'}} - \varepsilon_{\mathbf{k}})^2} + \cdots . \qquad (3)$$

Since we are mainly concerned with the region where the denominator vanishes, $V_{\mathbf{kk'}}$ may be taken to be a constant. When one replaces the summations by integrals using the density of states $\rho(\varepsilon)$, one finds that the second term is

$$V^2\rho^2 \int_{\varepsilon_F}^{\varepsilon_F+D} d\varepsilon' \int_o^{\varepsilon_F} d\varepsilon \frac{1}{(\varepsilon'-\varepsilon)^2} , \qquad (4)$$

where ε_F is the Fermi energy and D is the cut-off of the electron energy (of the same order as ε_F). This is logarithmically divergent, because $\varepsilon'-\varepsilon$ can be as small as zero.

However, such a divergence of the normalization constant is irrelevant, if one is interested in an average value such as

$$\langle A\rangle = \langle\Psi|A|\Psi\rangle/\langle\Psi|\Psi\rangle.$$

In fact, when one calculates the above quantity with the potential as an expansion parameter, one finds that it is an expansion in terms of V/ε_F, which behaves quite normally. This means that the characteristic energy for the electrons is ε_F. If the small excitation energy comes into play, one may ask what is the energy scale for such low excitation. However, the fact is that the density of excitation modes increases continuously with the excitation energy, until it reaches $\sim\varepsilon_F$.

3. DYNAMICAL PERTURBATION IN METALS

It turns out that the divergence in question, such as (4), does come into play in physical problems, when the perturbation acting on the electrons is dynamical (or time-dependent). The simplest example is the case where the potential $V(r)$ is zero for $t<0$ and is suddenly switched on at $t=0$. For $t<0$, the wave function is Φ_o (the Fermi sphere), whereas for $t>0$ it is Ψ, (2). The overlap integral $\langle\Psi|\Phi_o\rangle$ occurs in physically observable quantities in this case. From (2), one finds

$$\langle\Psi|\Phi_o\rangle = 1/N, \tag{5}$$

which is zero because N is divergent. This is called Anderson's orthogonality theorem. We shall see later how quantities like $\langle\Psi|\Phi_o\rangle$ are related to physical quantities. We mention here three actual cases, where a dynamical perturbation acts on the metal electrons.

3.1 Magnetic Alloys [1]

When the metal contains magnetic atoms as impurities, they exert a spin-dependent potential on the electrons. Since the spins of the magnetic atoms fluctuate, the electrons feel a time-dependent potential. Physical quantities, such as the specific heat, magnetic susceptibility and electrical resistivity of magnetic alloys show anomalous temperature dependence, which originates from the divergence mentioned.

3.2 X-Ray Spectra of Metals [2]

When an X-ray photon excites an inner electron to an empty level, a positive hole is left behind. It may be alive for a short time, especially for soft X-ray absorption. The hole will exert a potential $V(r)$ on the electrons. The wave functions before and after the photon absorption will be Φ_o and Ψ, and the absorption coefficient involves the overlap integral $\langle\Psi|\Phi_o\rangle$. The result of this is a singularity at the edge of the absorption spectrum. Similar singularity also occurs for X-ray emission in metals.

3.3 Diffusion of Muons or Protons in Metals [3]

Muons or protons sit in interstitial sites in metals and may diffuse among many sites. They have a positive charge and exert a potential on the electrons. In this case, the center of the potential moves. In this sense, it is a dynamical perturbation on the electrons. We shall see in this talk what effect the electrons have on the diffusion constant of these particles.

4. THE MODEL

When niobium metal, for example, is exposed to H_2 gas, the H_2 molecules become attached to the metal surface and dissociate into H atoms, which then diffuse into the body of the metal.

Positive muons, after injection into metals, are believed to lose the kinetic energy immediately and diffuse with thermal energy before they decay within the life time of 2.2μs. They are regarded as an isotope of the proton.

In order to diffuse among sites, they must jump over the potential hill, so the diffusion constant will decrease rapidly as the thermal energy of these particles decreases below the height of the hill. This is actually observed for the diffusion of atoms, including heavier atoms. At helium temperatures, heavier atoms diffuse very slowly and are practically fixed to one of the sites. Even in this situation, muons or protons may diffuse by tunneling. We are concerned here with this tunneling and shall investigate the effect of the metal electrons on the tunneling.

In order to investigate such effects, we first set up a simple model (Fig. 1): (i) We consider only two sites for the particle. (ii) Among levels in each potential well, we consider only the lowest ones. We denote their wave functions as $\chi_1(\mathbf{R})$ and $\chi_2(\mathbf{R})$, respectively, where $\mathbf{R}$ is the coordinate of the particle. (iii) The tunneling matrix between χ_1 and χ_2 is denoted by Δ. Then the eigenstates will be

$$\frac{1}{\sqrt{2}}(\chi_1 \pm \chi_2), \tag{6}$$

with the energy ±Δ. (iv) The interaction between the particle and the electrons is denoted by $\sum_i V(\mathbf{r}_i - \mathbf{R})$ where $\mathbf{r}_i$ is the electron coordinate.

One might treat this problem in the adiabatic scheme: One would first fix the particle at $\mathbf{R}$ and solve for the electron wave function, which involves $\mathbf{R}$ as a parameter. One would then find the adiabatic potential for the particle and solve for the particle wave function with this potential added to the bare potential. In this scheme, the effect of the electrons is involved only in the adiabatic potential. It turns out , however [4], that this scheme involves a divergence closely related to (4). This means the electrons are not fast enough to follow the particle adiabatically. This is because of the low excitation energy characteristic

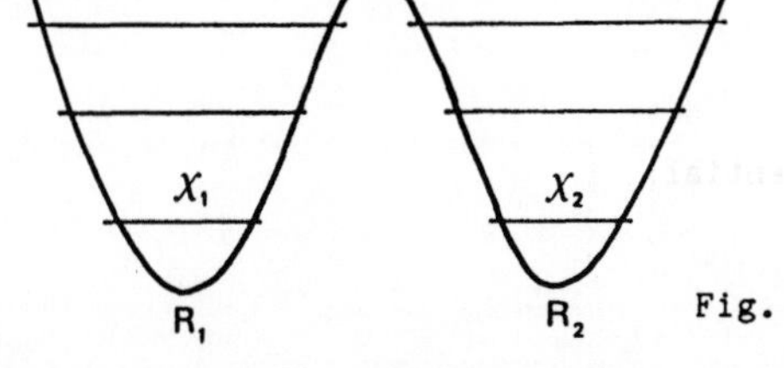

Fig. 1. Model potential for muons

of the metal electrons. Due to this non-adiabatic effect, the electrons do not see the potential centered at the instantaneous position of the particle, but see some time-averaged potential. This amounts to a change of item (iv) into (iv'): When the particle is in the state χ_1, the electrons see the potential

$$\sum_i V(\mathbf{r}_i - \mathbf{R}_1),$$

where $\mathbf{R}_1$ is the center of the wave function χ_1. Similarly, an effective potential is introduced when the particle is in χ_2. The intensity of the potential may be reduced somewhat because of the averaging.

Having set up the model, we can now make a standard treatment of it. For example, we may calculate the partition function:

$$Z = \mathrm{Tr}\, e^{-\beta H}, \tag{7}$$

where $\beta = 1/k_B T$ and H is the Hamiltonian defined by (i)-(iv'). Before doing so, we discuss qualitative feature of our model. We first place the particle in χ_1. The electron wave function will be obtained similarly to (2), the difference being that the potential is now centered at $\mathbf{R}_1$. Thus, the wave function is expressed by

$$\Psi_1 = \frac{1}{N}\left[\Phi_o - \sum \frac{e^{i(\mathbf{k}-\mathbf{k}')\cdot\mathbf{R}_1} V_{\mathbf{kk}'}}{\varepsilon_{\mathbf{k}'} - \varepsilon_{\mathbf{k}}} \Phi_{\mathbf{k}+\mathbf{k}'} + \cdots\right]. \tag{8}$$

The total wave function is $\chi_1\Psi_1$. Similarly, when the particle is placed in χ_2, the total wave function will be $\chi_2\Psi_2$, where Ψ_2 is expressed by (8) with $\mathbf{R}_1$ replaced by $\mathbf{R}_2$. Then, the effective tunneling matrix will be

$$\Delta_{eff} = \langle\chi_2\Psi_2|H'|\chi_1\Psi_1\rangle = \langle\chi_2|H'|\chi_1\rangle\langle\Psi_2|\Psi_1\rangle = \Delta\langle\Psi_2|\Psi_1\rangle, \tag{9}$$

where H' is a Hamiltonian which causes the tunneling of the particle. Thus, it is renormalized by the overlap integral between the initial and final electron states.

From (8) and a similar expression for Ψ_2, we find

$$\langle\Psi_2|\Psi_1\rangle = 1 - \sum \frac{[1 - e^{i(\mathbf{k}-\mathbf{k}')\cdot(\mathbf{R}_1-\mathbf{R}_2)}]\cdot|V_{\mathbf{kk}'}|^2}{(\varepsilon_{\mathbf{k}} - \varepsilon_{\mathbf{k}'})^2} + \cdots . \tag{10}$$

One may take the angular average of the exponential:

$$\langle\Psi_2|\Psi_1\rangle = 1 - \left(1 - \frac{\sin^2 k_F |\mathbf{R}_1-\mathbf{R}_2|}{k_F^2 |\mathbf{R}_1-\mathbf{R}_2|^2}\right) V^2 \sum_{\substack{k<k_F \\ k'>k_F}} \frac{1}{(\varepsilon_{\mathbf{k}'}-\varepsilon_{\mathbf{k}})^2} + \cdots. \quad (11)$$

When one replaces the summations by integrals, one has a logarithmic divergence. When one keeps the discreteness of the levels, it is found that the sum is expressed by $\rho^2 \log N_e$ + const., where N_e is a macroscopic number proportional to the size of the system. It may be the total number of lattice points in the metal or the total number of the electrons. On the other hand, the constant in the above expression is independent of the size of the system. We neglect this term. Thus we define a dimensionless parameter K:

$$K = V^2\rho^2 \left(1 - \frac{\sin^2 k_F |\mathbf{R}_1-\mathbf{R}_2|}{k_F^2 |\mathbf{R}_1-\mathbf{R}_2|^2}\right), \quad (12)$$

and write

$$\langle\Psi_2|\Psi_1\rangle = 1 - K\cdot\log N_e + \cdots.$$

Higher-order terms are also calculated [5] and one finds

$$\langle\Psi_2|\Psi_1\rangle = \exp(-K\cdot\log N_e). \quad (13)$$

This tends to zero as the size of the system tends to infinity.

At finite temperatures, Φ_o in (8), for example, must be an excited state. In this case, the overlap integral turns out to be

$$\langle\Psi_2|\Psi_1\rangle = \exp[-K\cdot\log(\varepsilon_F/k_BT)] = \left(\frac{k_BT}{\varepsilon_F}\right)^K . \quad (14)$$

Having obtained Δ_{eff}, we have a simple picture for our system: The wave function is expressed by

$$(\chi_1\Psi_1 \pm \chi_2\Psi_2)/\sqrt{2} \quad (15)$$

with the energy $\pm\Delta_{eff} = \pm\Delta(k_BT/\varepsilon_F)^K$. The effect of the electrons on the tunneling particle is to renormalize the tunneling matrix by a factor which is proportional to a power of the temperature.

5. HOPPING RATE

Let us consider the hopping rate of the particle from one site to the other. We assume that the state at t=0 is represented by $\chi_1\Psi_1$. What is the probability that the particle is found in χ_2 at time t? Call this probability $P_{12}(t)$. Then we speculate

$$P_{12}(t) = \sin^2 \Delta_{eff} t. \tag{16}$$

If the states (15) were exact, this would be correct. However, they have a finite life time τ_c due to the effect of the interaction V which is not taken into Δ_{eff}. We distinguish two cases:

(a) $\tau_c \gg \Delta_{eff}^{-1}$

Then, (16) is nearly correct and $P_{12}(t)$ shows a damped oscillation (Fig. 2).

(b) $\tau_c \ll \Delta_{eff}^{-1}$

Then, $P_{12}(t)$ will increase quadratically for small t until the coherent increase is perturbed at $t \sim \tau_c$. From this time on, P_{12} again increases quadratically until $t \sim 2\tau_c$, and so on (Fig. 3). The net result is a linear increase $P_{12} = Wt$, with $W = \Delta_{eff}^2 \tau_c$ (Fig. 4). W is defined as a hopping rate and is related to the diffusion constant D by $D = \alpha W a^2$, if there are regular array of the sites. a is the distance between nearest sites, and α is a constant of the order of unity.

Most of the practical cases seem to belong to case (b), so that the diffusion constant in the tunneling regime of these cases will be given by $D = \alpha W a^2$. Let us then

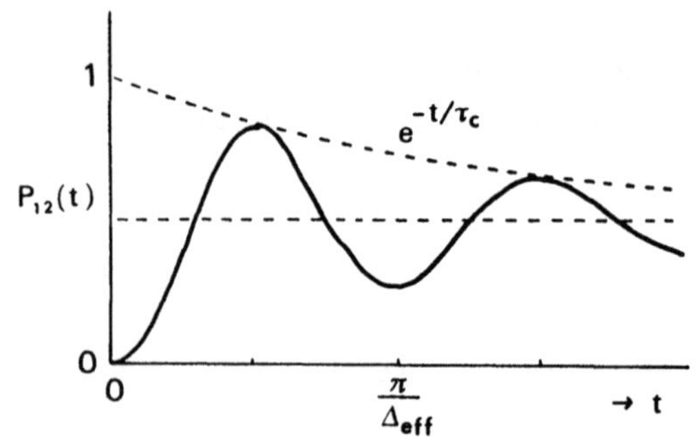

Fig. 2. $P_{12}(t)$ for weak damping

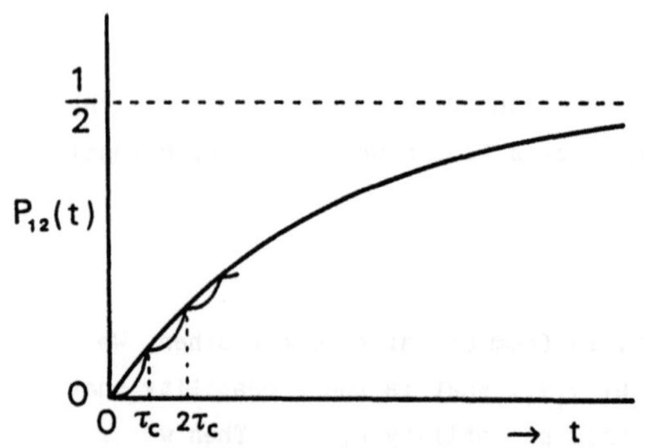

Fig. 3. $P_{12}(t)$ for strong damping

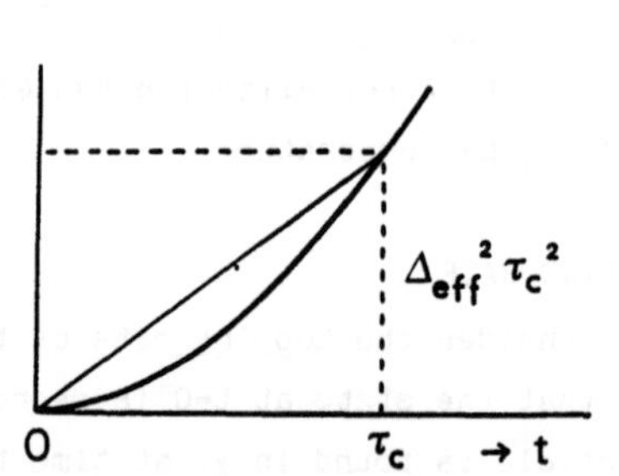

Fig. 4. Close-up view of Fig. 3

ask what is τ_c. It will be the inverse of the level broadening of states (15). The broadening comes from the interaction with the conduction electrons, and so is expected to be proportional to k_BT in the same way as the Korringa broadening. (The Zeeman level of a spin in metals has a broadening proportional to k_BT, which arises from the interaction with the metal electrons. This is called Korringa broadening.) As a result, we expect that $D \propto T^{2K-1}$. YAMADA, SAKURAI and TAKESHIGE [5] obtained an expression for K which involves higher order terms of V, and showed that K is always less than one half. Thus, D increases as the temperature decreases.

Let us make a more concrete calculation. We represent the particle in terms of the Pauli matrix σ_z: $\sigma_z=1$ means the particle in χ_1 and $\sigma_z=-1$ that in χ_2. The tunneling Hamiltonian is then represented by $\Delta\cdot\sigma_x$. The total Hamiltonian is $H+\Delta\cdot\sigma_x$, where

$$H = \frac{1}{2}(1 + \sigma_z)\cdot H_1 + \frac{1}{2}(1 - \sigma_z)\cdot H_2 , \tag{17}$$

$$H_1 = -\frac{\hbar^2}{2m}\sum_i \Delta_i + \sum_i V(\mathbf{r}_i - \mathbf{R}_1) , \tag{18}$$

$$H_2 = -\frac{\hbar^2}{2m}\sum_i \Delta_i + \sum_i V(\mathbf{r}_i - \mathbf{R}_2) . \tag{19}$$

When the system is in $\chi_1\Psi_1$ at t=0, the state at t is given by

$$\Phi(t) = e^{-i(H + \Delta\cdot\sigma_x)t}\chi_1\Psi_1 . \tag{20}$$

Here, χ_1 is not a function of $\mathbf{R}$, but is represented by

$$\begin{pmatrix} 1 \\ 0 \end{pmatrix}$$

and similarly for χ_2.

We expand (20) in terms of Δ:

$$\Phi(t) = e^{-iHt}\left[1 + \sum_n (-i\Delta)^n \int\cdot\cdot\int_{t>t_n>\cdot\cdot>t_1>0} \sigma_x(t_n)\cdot\cdot\sigma_x(t_1)\, dt_1\cdot\cdot dt_n\right]\chi_1\Psi_1 . \tag{21}$$

$\sigma_x(t)$ is the interaction representation of σ_x. Then, the probability that the particle is found in χ_2 at t is expressed by

$$P_{12}(t) = \sum_i |\langle\chi_2\Psi_i|\Phi(t)\rangle|^2$$

$$= \sum_{n,m} (-)^{n+m-1}\,\Delta^{2(n+m)} \int\cdot\cdot\int_{\substack{t>t_{2n}>\cdot\cdot>t_1>0 \\ t>\tau_{2m}>\cdot\cdot>\tau_1>0}} \langle\sigma_x(t_{2n})\cdot\cdot\rangle\, dt_1\cdot\cdot d\tau_{2m} . \tag{22}$$

Here, only even powers of Δ remain.

A sophisticated calculation [6] gives us the answer for the average in the above expression. It turns out that there is an energy scale characteristic to our system:

$$\tilde{\Delta} = [4\cos\pi K\cdot\Gamma(1-2K)]^{\frac{1}{2-2K}}\cdot\Delta\cdot\left(\frac{\Delta}{\varepsilon_F}\right)^{\frac{K}{1-K}}, \tag{23}$$

where Γ is the gamma function. We distinguish two cases:

(A) $k_BT \gg \tilde{\Delta}$

Here, we find

$$P_{12}(t) = (1 - e^{-2Wt})/2, \qquad (\simeq Wt \quad \text{for small } t) \tag{24}$$

where

$$W = \frac{1}{\sqrt{\pi}}\frac{\Gamma(K)}{\Gamma(\frac{1}{2}+K)}\frac{\Delta^2}{k_BT}\left(\frac{\pi k_BT}{\varepsilon_F}\right)^{2K}. \tag{25}$$

This is proportional to T^{2K-1} as we have expected. If we write $W=\Delta_{eff}^{\,2}\tau_c$ with $\Delta_{eff}=\Delta(\pi k_BT/\varepsilon_F)^K$, we find $\tau_c^{-1}\propto[\Gamma(\frac{1}{2}+K)/\Gamma(K)]\cdot k_BT$. One then observes that $k_BT\gg\tilde{\Delta}$ implies $\Delta_{eff}\ll\tau_c^{-1}$. This case corresponds to the case (b) mentioned previously.

(B) $T=0$

One finds

$$2P_{12}(t) = 1 + \frac{1}{2\pi i}\int_{-\sigma-i\infty}^{-\sigma+i\infty}\frac{e^{pt}}{p\left[1+\left(-\frac{\tilde{\Delta}}{p}\right)^{2-2K}\right]}\,dp, \tag{26}$$

where $\sigma>0$. This is a known function and represents a damped oscillation plus a power-law decay.

The Fourier transform of (26) will be easier to understand. It has a peak at $\hbar\omega\simeq\cos\pi K\cdot\tilde{\Delta}$ with the width $\sim\sin\pi K\cdot\tilde{\Delta}$. For small K, this represents a well-defined peak. Even in this case, the peak will be broadened as the temperature goes above $\tilde{\Delta}/k_B$ (case (A)).

6. COMPARISON WITH EXPERIMENTS

With the theoretical prediction in mind, let us discuss two cases of related experiments. The first one is diffusion of the positive muon in metals. As mentioned previously, there are regularly arrayed equivalent interstitial sites in metals and the muon diffuses among them by tunneling at low temperatures. When $k_BT\gg\tilde{\Delta}$, one may consider just two neighbouring sites and discuss diffusion in terms of the hopping rate between them. This means that successive hoppings are not correlated. When $k_B T \ll \tilde{\Delta}$, however, the muon may tunnel among many sites coherently. Then, the model defined by (i)-(iv') is not applicable to this case, where one must consider many sites. The case of the positive muon in copper belongs to the case $k_B T \gg \tilde{\Delta}$. Fig. 5 shows experimental data for the hopping rate W of this case as a function of the temperature [7]. As the temperature is decreased from room temperature, W decreases rapidly. This is due to a process of the activation type. However, the actual process is not as simple as jumping over the potential hill, but is believed to be the "underbarrier tunneling". At T~50K, the W vs T curve shows a minimum, which means the tunneling process has now set in. Below this temperature the hopping rate is determined by the interaction with the metal electrons as well as that with the phonons. The effect of the latter mechanism decreases rapidly [8] until at T ~10K, where the interaction with the electrons becomes the dominant mechanism determining the hopping rate. The theoretical curves of Fig. 5 take the three mechanisms into account: the "underbarrier tunneling", the interaction with the phonons and that with the electrons. Independently from parameters related to the former two mechanisms, we can determine those for the electron interaction by fitting the curve to data below 10K. Our theory (25) involves three parameters, ε_F, Δ and K. From the slope of the curve, we obtain the values of K. ε_F may be taken to be ~5eV. Then, from the absolute value of W, we obtain that of Δ. These values are shown in the figure. Using these values in (23), we find $\tilde{\Delta} \sim 10^{-3}$K, so this certainly corresponds to case (A) at practical temperatures.

The experimental points level off at the lowest temperatures. One possible mechanism for this will be noted: Natural copper is a mixture of isotopes, Cu^{63} and Cu^{65}. The muon level at each interstitial site may depend on the distribution of

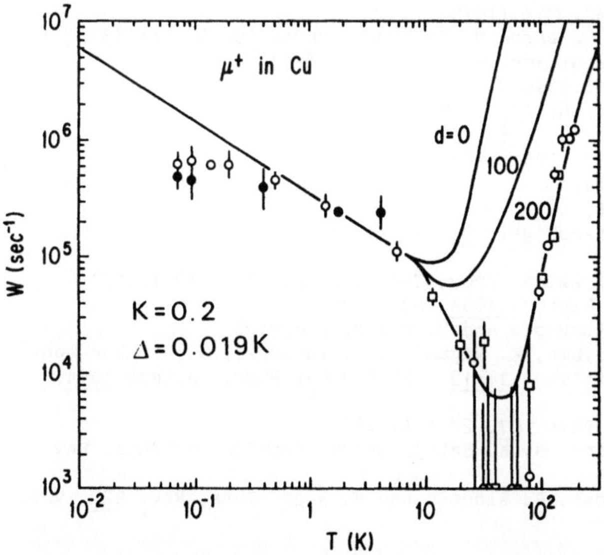

Fig. 5. The hopping rate of the positive muon in copper

isotopes among the six atomic positions surrounding the site. This is because the muon rather strongly distorts the surrounding atomic positions to be in a more stable state, so that dynamics of the atoms has an effect on the muon level. This causes fluctuation of the muon levels. As $k_B T$ becomes smaller than the level fluctuation due to this mechanism (or other fluctuations), the hopping rate may level off.

The other experiment related to the theory developed here is neutron spectroscopy of the particle levels. The protons dissolved in niobium at high temperatures precipitate as the temperature goes down to helium temperature. When niobium contains a small amount of impurity oxygen in it, the proton is trapped around an oxygen atom and remains in that position even at very low temperatures. This proton is believed to constitute a typical two-level system which is represented by the model defined by (i)-(iv'). In the inelastic neutron scattering experiment, the proton may be raised from the state $\chi_1\Psi_1+\chi_2\Psi_2$ to $\chi_1\Psi_1-\chi_2\Psi_2$ or the reverse of it. At $k_B T \ll \tilde{\Delta}$ one may observe an isolated peak, whereas at $\kappa_B T \gg \tilde{\Delta}$, the level broadening τ_c^{-1} is larger than the level separation Δ_{eff}. Such a cross-over is actually observed for the system mentioned above [9]. At T=0.25K, one observes a well-defined inelastic peak at the energy 0.2mK, whereas at T=1.44K the peak is already very flat. Such a strong temperature dependence can only be explained in terms of the interaction with the metal electrons.

References

1. J. Kondo: Solid State Physics 23, 183 (1969)
   Magnetism V, ed. by H. Suhl, (Academic Press 1973)
   G. Grüner and A. Zawadowski: Prog. Low Temp. Phys. VIIb, ed. by D. F. Brewer, (North-Holland 1978) p.592
   K. G. Wilson: Rev. Mod. Phys. 47, 773 (1975)
   N. Andrei, K. Furuya and J. H. Lowenstein: Rev. Mod. Phys. 55, 331 (1983)
   A. M. Tsvelick and P. G. Wiegmann: Adv. Phys. 32, 453 (1983)
2. G. D. Mahan: Phys. Rev. 163, 612 (1967)
   P. Nozieres and C. T. de Dominicis: 178, 1097 (1969)
3. J. Kondo: Physica 125B, 279 (1984); 126B, 377 (1984)
   K. Yamada: Prog. Theor. Phys. 72, 195 (1984)
4. J. Kondo: Physica 124B, 25 (1984)
5. J. Kondo: Physica 84B, 40 (1976)
   K. Yamada, A. Sakurai and M. Takeshige: 70, 73 (1983)
6. J. Kondo: in reference [3]
   K. Yamada, A. Sakurai and S. Miyazima: Prog. Theor. Phys. 73, 1342 (1985)
   U. Weiss and H. Grabert: Phys. Letters 108A, 63 (1985)
   H. Grabert, S. Linkwitz, S. Dattagupta and U. Weiss: preprint
7. R. Kadono, J. Imazato, K. Nishiyama, K. Nagamine, T. Yamazaki, D. Richter and J. -M. Welter: Hyperfine Interactions 17-19, 109 (1984); Phys. Letters 109A, 61 (1985)
8. Y. Kagan and M. I. Klinger: J. Phys. C7, 2791 (1974)
9. H. Wipf, A. Margel, S. M. Shapiro, S. K. Satija and W. Thomlinson: Phys. Rev. Letters 46, 947 (1981)
   A. Margel, J. J. Rush, J. M. Rowe, D. Richter and H. Wipf: Phys. Rev. B27, 927 (1983)

Mean Field Theory for Spin Glasses

G. Parisi

Dipartimento di Fisica, II Universitá di Roma
"Tor Vergata" and INFN, Sezione di Roma, Roma, Italy

In recent years much progress has been made in this field, and after a great deal of effort self-consistent mean field approximation has been obtained [1] (sometimes this construction goes under the name of broken replica theory). This theory should be exact for weak long-range forces or when the dimensions of the space become very large (fluctuations are neglected). The ideas involved in this construction are rather different from those of the mean field approximation for other models and take care of the peculiar properties of spin glasses. In this paper we will not show how these ideas have been developed. We will also skip most of the technical details and concentrate our attention on a few important physical results.

A spin glass is a disordered magnet; in a typical example 50% of the randomly chosen bonds between spin pairs are ferromagnetic while the others are antiferromagnetic. From the experimental point of view [2], spin glasses are characterized by the presence of a low-temperature phase in which the relaxation time for attaining equilibrium is very large, as is the case in real glasses. The magnetic susceptibility in the presence of a time dependent magnetic field depends on the frequency, even in the region of a few Hertz and the power spectrum of the magnetization behaves as the inverse of the frequency (1/f noise).

From the theoretical point of view, these properties may be connected to the practical impossibility of finding numerically the ground state of a spin glass, if the number of spins is not very small; indeed, according to conventional wisdom, the number of operations which any algorithm needs to find the ground state increases exponentially with the volume (the problem is NP complete in more than 2 dimensions). There is an exponentially large number of spin configurations which are local ground states, in the sense that their energy increases if only one spin is flipped. It is very easy to find efficient algorithms for finding local minima. In contrast, it is very difficult for the algorithm to find the global minimum, i.e. the true ground state.

Exhaustive numerical simulations on small systems show that different minima (of the energy at zero temperature or of the free energy at finite temperature) are macroscopically different and that the energy difference between macroscopically different minima remains of O(1) even when the volume goes to infinity. This situation is rather unusual in statistical mechanics; macroscopically different configurations with the same total free energy are normally present at the point where a first order phase transition happens (which is an isolated point). For spin glasses this phenomenon is present in a large range of temperatures and magnetic fields where no first order phase transition is seen from the thermodynamical point of view. A strong effort has been needed to find the appropriate theoretical framework to study these systems.

The origin of this peculiar behaviour is the presence of frustration; it is not possible that all pairs of spins which are connected by a ferromagnetic (antiferromagnetic) bond have the same (the opposite) sign. Different terms in the Hamiltonian push in different directions and the number of possible compromises is very large. For example in the Ising case the Hamiltonian is

$$H = -\frac{1}{2} \sum_{i,k} J_{i,k}\sigma_i\sigma_k - h\sum_i \sigma_i, \tag{1}$$

where h is the magnetic field and the J's are the coupling between the spins. It is clear that if the product $J_{12}J_{23}J_{31}$ is negative, it is impossible to find configurations of the σ's such that $\sigma_1 J_{12}\sigma_2$, $\sigma_2 J_{23}\sigma_3$, $\sigma_3 J_{31}\sigma_1$ are all positive.

There are many systems in which different terms in the Hamiltonian, or different constraints, are in competition with one another. This happens very frequently when the system is complex. Typical examples are real glasses, other NP complete problems like the travelling salesman or the matching problem, protein folding, biological organization, prebiotic evolution, neuron network and so on. At present, many of the concepts that have been developed in the study of spin glasses, are also becoming useful for these systems, and it is quite possible that, in the long run, the applications of these ideas beyond solid state physics will be the most interesting ones. As we shall see later, the charateristic feature of spin glasses in the low temperature phase is the existence of many stable equilibrium states, which differ microscopically, but not macroscopically, from one another. The theory of these systems will not be simple [3], because it must reflect the complexity of the problem we face.

Suppose that we have a system whose Hamiltonian $H_J[\sigma]$ depends on the configuration $[\sigma]$ of the system and on some control variables J, which are distributed according to a probability distribuion P[J]. For each choice of the J's we can compute the partition function

$$Z_J = \sum_{\{\sigma\}} \exp\{-\beta H_J[\sigma]\} \tag{2}$$

and free energy density

$$F_J = -\frac{1}{\beta N}\ln Z_J = -\frac{1}{\beta N}\ln\{\sum_{\{\sigma\}}\exp\{-\beta H_J[\sigma]\}\}, \tag{3}$$

where N is the total number of variables (i.e. of σ's).

Standard statistical mechanics deals with the problem of computing F for a given J. Here we suppose that the J's are not known, but that they are random variables and that their probability distribution is known. The J's are called quenched variables. Physically, the necessity of computing the free energy F at fixed J's results from the fact that changes of the J's happen on a time scale which is infinitely larger than the time scale characterizing the changes in the σ's.

If the form of the interaction is not too pathological, standard thermodynamic arguments imply that in the limit of N going to infinity (infinite volume limit) the free energy density assumes the same value for each set of the control parameters (J's) that has a non-vanishing probability.

We are interested in computing the average value of the free energy density:

$$F = \sum_J P[J]F_J \equiv \overline{F_J}. \tag{4}$$

The direct computation of (4) is not simple because it is not of the form which is usual in statistical mechanics; for example the computation of $\ln\{\sum_J P[J] \times \sum_{\{\sigma\}}\exp\{-\beta H_J[\sigma]\}\}$, the so called annealed case, would be more straightforward. Here we first choose the J's and then compute the probability distribution for the σ's; the variables J's are called quenched.

The replica method was originally proposed as a trick to simplify the computation of (4). Later it was found that the replica method (when the replica symmetry is broken) is very powerful in coding, in a simple and compact way, quite complex properties of the system.

In a nutshell the idea at the basis of the replica method is very simple. We start with some preliminary definitions;

$$Z_n = \sum_J P[J]\{Z_J\}^n \equiv \overline{(Z_J)^n} \tag{5}$$

$$F_n = -\frac{1}{\beta n N}\ln Z_n.$$

Using the relations $\sum_J P[J]=1$ and $A^n \approx 1+n\ln(A)$ for $n\approx 0$, it is evident that

$$\lim_{n\to 0} F_n \equiv F_0 = \overline{F_J}\ . \tag{6}$$

For integer n we can write;

$$(Z_J)^n = \prod_{a=1}^{n}\sum_{\{\sigma^a\}}\exp\{-\sum_{a=1}^{n}\beta H[\sigma^a]\}, \tag{7}$$

where we have introduced n replicas of the same system. Indeed the partition function of n replicas of the same system is the partition function of the original system to the power n. The variables σ_i^a carry two indices; the upper one denotes the replica and runs from 1 to n and the lower one denotes the site and runs from 1 to N. At the end we must take the two limits $n\to0$ and $N\to\infty$, which are supposed to commute.

The replica trick consists of the following steps; we use (7) to define the function F_n for integer n, extend this function to an analytic function of n, and finally compute $F_0=F$. If the partition function is expanded in powers of β then F_n is a polynomial in n and the method is quite safe. Problems will arise in the low temperature region.

We can apply this method to the case of infinite range spin glasses where the J's are independent Gaussian random variables with zero mean and variance 1/N:

$$\overline{J_{i,k}}=0, \quad \overline{J_{i,k}^{\,2}}=\frac{1}{N}, \quad J_{i,k}=J_{k,i}. \tag{8}$$

The thermodynamic limit is obtained when N becomes infinity. The factor 1/N in (8) has been chosen in such a way that at fixed β the total energy is proportional to N and therefore the energy density is N-independent.

For fixed J's we expect that in the high-temperature phase the local magnetizations $m_i\equiv\langle\sigma_i\rangle$ are different from zero at non-zero magnetic field and they become zero when the magnetic field goes to zero. On the other hand, we naively expect that in the low temperature region there should be some freezing of the spins in the position which is most energetically favourable, hence the m_i should be different from zero also at h=0. It is evident that m_i depends on the J's and it will sometimes be positive and sometimes negative (a detailed computation shows that $\overline{m_i}=0$ at h=0) so it is convenient to characterize the system in terms of the quantity

$$q_{EA} \equiv \frac{1}{N}\sum_{i=1}^{N} m_i^{\,2}, \tag{9}$$

the so called Edwards-Anderson order parameter [4]. It is possible to show that using a good definition of the m's, q_{EA} is also equal to $\overline{m_i^2}$. Summarizing, q_{EA} should be equal to zero at h=0 for temperatures greater than the critical temperature (T_c), while it should be different from zero at low temperatures.

If we apply (5) to the infinite-range case then, after some simple computations, we find that

$$Z_n = \int d[Q]\exp\{-NA[Q]\} \tag{10}$$

$$A[Q] = -\frac{n\beta^2}{4}+\frac{1}{4}\sum_{a=1}^{n}\sum_{b=1}^{n} Q^2_{ab} - \ln Z[Q]$$

$$Z[Q] = \sum_{\{S\}} \exp\{-\beta H[Q,S]\}$$

$$H[Q,S] = \sum_{a=1}^{n} \sum_{b=1}^{n} Q^2_{ab} S_a S_b - h \sum_{a=1}^{n} S_a$$

$$F_n = -\frac{1}{\beta} \min A[Q] \quad \Rightarrow \quad \frac{\partial A}{\partial Q_{ab}} = 0,$$

where the matrix Q is an $n \times n$ symmetric matrix, zero on the diagonal. The integral is done on all non-diagonal elements of the matrix Q and goes from $-\infty$ to $+\infty$, the sum over $\{S\}$ runs over the 2^n configurations of the variables S_a, $a=1,..,n$ ($S_a = \pm 1$).

The function A[Q] is left invariant when we exchange some of the columns (or the rows) of the matrix Q, and therefore the group of permutations of n elements (P_n or S_n) is a symmetry of the problem (all replicas are equivalent!). This group is often called the replica group.

For positive n, the minimum of A can be found and the matrix Q has the following form

$$Q_{a,b} = q, \ \forall a,b \ a \neq b \ (Q_{a,a} = 0, \ \forall a). \tag{11}$$

This form of the matrix Q is the only one which is left invariant by the action of the replica group and it is therefore the natural solution, usually named the replica symmetric solution.

If we analytically continue the solution of the equation $dA/dq=0$, up to $n=0$, we find the following equation for q

$$q = \int dz/(2\pi)^{1/2} \exp(-z^2/2) \mathrm{th}^2(\beta q^{1/2} z + \beta h), \tag{12}$$

where the integral over z goes from $-\infty$ to $+\infty$.

At zero magnetic field (12) has only the solution $q=0$ for $1/\beta \equiv T > T_c = 1$; for $T < T_c$ there is another solution (the physical one) which is different from zero. In conclusion, there is a phase transition at $T=1$, $h=0$ while there is no transition at $h \neq 0$.

Everything seems perfect, but unfortunately a detailed computation shows that the entropy (which in a discrete system is non-negative by definition) becomes negative at low temperature and at zero temperature we get $S(0) = -1/(2\pi) \approx -.17$. In this case the replica method leads to a disaster!

It is evident that the saddle-point method can be applied correctly only if the eigenvalues of the Hessian matrix($\partial^2 A/\partial Q \partial Q$) are non-negative, as can be checked by computing the correction to the saddle-point result. A careful computation [5] shows that for any value of the magnetic field the Hessian, evaluated at the solution of (12), becomes negative at sufficient low temperature when $n<1$. The ansatz (11) does not give the correct saddle point for evaluating the integral. At low temperatures we must look for a new solution of the equation

$$\partial A/\partial Q = 0, \qquad (13)$$

whose Hessian has no negative eigenvalues.

The naive replica approach fails because there is a phase transition in the region $0<n<1$; we cannot do the analytic continuation for the free energy, but we can perform the analytic continuation of the equations which determine the matrix Q and look for other solutions of the equation at $n=0$.

Unfortunately, if the matrix Q does not have the form shown in (11), it is not left invariant by the action of the replica group and we have to specify in detail the value of the different elements of the matrix. Since the matrix Q has no element in the limit n going to zero, the meaning of this operation is not clear a priori; moreover it is not clear at this stage why the correct saddle point should not be left invariant by the action of the replica group and what are the physical consequences of such non-invariance.

On the other hand, the value of many thermodynamic functions computed within the above replica symmetric solution do not disagree too much with the numerical data. Superficially it seems that only small modifications of the approach are needed to get the correct results.

In order to avoid inconsistencies we will always write formulae which are correct for integer n and make the analytic continuation at $n=0$ only at the end. In other words, we will consider matrices $q^{(n)}_{a,b}$ that depend on n in a simple way, compute all the quantities we need for integer n, and only at the end we will perform the continuation of the results at $n=0$.

From this point of view the 0×0 matrix $q^{(0)}_{a,b}$ is defined in terms of the matrices $q^{(n)}_{a,b}$ for all values of n and therefore the space of all 0×0 matrices is an infinite dimensional space.

To find the physically relevant saddle point, we must proceed by trial and error. The space of 0×0 matrices is very large and it is impossible at the present moment to write a workable expression for a generic element of such a space.

At present, only one reasonable form of the matrix is known. There is a widespread agreement that this choice is the correct one because the results are very satisfactory. It is quite possible that for systems other than spin glasses (e.g. for real glasses), different choices of the matrix $q_{a,b}$ are required.

A natural suggestion for the matrix Q consists of dividing the n replicas into n/m groups of m replicas. (Of course n must be multiple of m, i.e. n/m must be an integer). We set $Q_{a,b}=q_0$, if a and b belong to the same group, and $Q_{a,b}=q_1$, if a and b belong to different groups (we do not consider the $Q_{a,a}$'s which are identically equal to 1). In other words

$$Q_{a,b} = q_0 \quad \text{if} \quad I(a/m) = I(b/m)$$

$$Q_{a,b} = q_1 \quad \text{if} \quad I(a/m) \neq I(b/m), \tag{14}$$

where $I(x)$ is an integer-valued function; its value is the smallest integer which is greater than or equal to x.

Each column of the matrix has m-1 off-diagonal elements which are equal to q_0 and n-m which are equal to q_1. (The total number of off-diagonal elements is n-1, i.e. -1 in the limit $n\to 0$.) According to (14) we have in the limit $n\to 0$;

$$P(q) = m\delta(q-q_0) + (1-m)\delta(q-q_1), \tag{15}$$

where the function P(q) is defined as

$$P(q) = \frac{1}{n(n-1)/2} \sum_{a,b} \delta(q_{a,b}-q). \tag{16}$$

We will see later that the function P(q) must be non-negative and this is possible only if

$$0 \leq m \leq 1. \tag{17}$$

It is obvious that, if we exclude the two uninteresting cases m=0 and m=1, m cannot be an integer satisfying the inequality (17). On the other hand, the limit $n\to 0$ is obtained by doing an analytic continuation in m and nothing seems to forbid that in such a process m may take non-integer values.

If we evaluate the free energy we find after some computations that

$$F[Q] = F(q_0,q_1,m) = -\beta^2/4[1+mq_1{}^2+(1-m)q_0{}^2-2q_0{}^2]$$
$$-\int dp_{q_1}(z)m^{-1}\ln\{\int dp_{(q_0-q_1)}(y)\,\mathrm{ch}^m[\beta(z+y)]\}, \tag{18}$$

where we have used the short-hand notation for the Gaussian probability distribution with variance q and zero average;

$$dp_q(z) \equiv (2\pi q)^{-1/2}\exp(-z^2/2q)dz. \tag{19}$$

A simple computation shows that when m is set to 0 or to 1 we recover (as we should) the free energy of the replica symmetric approach (see (12) with $q=q_0$ or q_1 respectively.)

A careful analysis shows that the free energy A[Q] must be maximized with respect to all the variables, i.e. q_0, q_1 and m. If a maximum is found in the region 0<m<1 the results of this approach must be better than those obtained assuming unbroken replica symmetry; the true free energy takes values greater than those obtained from the unbroken replica approach.

The function F appearing in (18) can be easily computed numerically with high precision. Maximizing F with respect to the q_0, q_1 and m, one finds that in the whole region where the replica symmetry should be broken, i.e. at sufficient low temperatures, m is neither zero or one ($m \to 0$ both for $T \to 0$ and $T \to 1$ at zero magnetic field).

The properties of the solution are very satisfactory:

a) the zero temperature internal energy at zero magnetic field is -.7652 in good agreement with the numerical data (-.765 ±.01).

b) the zero temperature entropy at zero magnetic field has collapsed from $S(0) \approx -.16$ to $S(0) \approx -.01$.

c) a detailed computation of the Hessian shows that near the critical temperature at zero magnetic field the most negative eigenvalue is approximatively equal to $-C\tau^2/9$ (τ being equal to 1-T and C being a positive constant) where the same quantity in the replica symmetric case is equal to $-C\tau^2$. In other words the instability has decreased by a factor of 9.

We are clearly on the right track although we have not found the final answer.

The final form of the matrix Q is as follows: We introduce a set of integer numbers m_i ($i=0,\dots,k+1$) such that $m_0=0$ and $m_{k+1}=n$ and m_i/m_{i+1} is an integer (for $i=1,\dots,k+1$). We can divide the n replicas into n/m_k groups of m_k replicas, each group of m_k replicas is divided into m_k/m_{k-1} groups of m_{k-1} replicas and so on ... .

The off-diagonal elements of the matrix Q are thus given by

$$Q_{a,b} = q_i \text{ if } I(a/m_i) \neq I(b/m_i) \text{ and } \tag{20}$$

$$I(a/m_{i+1}) = I(b/m_{i+1}), \quad i = 0, \dots, k,$$

where the q_i's are a set of k+1 real parameters. For k=1 we recover the previous example and for k=0 we recover the unbroken replica symmetry theory.

An easy computation shows that

$$P(q) = \sum_{i=0}^{k} (m_i - m_{i+1}) \delta(q - q_i) \tag{21}$$

$$\lim_{n \to 0} \{ n^{-1} \sum_{a,b=1}^{n} Q_{a,b}^2 \} = - \sum_{i=0}^{k} (m_{i+1} - m_i) q_i^2 .$$

Equation (21) for P(q) is positive definite only if the m's satisfy the conditions

$$0 \leqslant m_{i+1} \leqslant m_i < 1. \qquad (22)$$

Detailed arguments suggest that we should look to the maximum of the free energy as a function of the q_i's and the m_i's. From now on we will assume that the conditions (22) are satisfied.

In order to keep track of the parameters q_i and m_i it is convenient to introduce the function $q(x)$ defined as

$$q(x) = q_i \text{ if } m_{i+1} < x < m_i. \qquad (23)$$

There is a one to one correspondence among the piecewise constant functions with k discontinuities and the parameters q_i and m_i (if (22) holds).

If the function $q(x)$ is a monotonic function, as we shall see it must be for physical reasons, the relation between the functions $P(q)$ and $q(x)$ is

$$dx/dq = P(q), \qquad (24)$$

where $q(x(q))=q$.

In the limit $k\to\infty$ the function $q(x)$ becomes arbitrary (any reasonable function can be approximated by a piecewise constant function) and

$$\lim_{n\to 0} \{n^{-1} \sum_{a,b=1}^{n} Q^k_{a,b}\} = - \int_0^1 dx\, q^k(x). \qquad (25)$$

In this formulation the free energy becomes a functional $F[q]$ of $q(x)$. It must be maximized with respect to $q(x)$ and in this way we find the correct solution of the infinite-range model.

The replica formalism is rather far removed from physical intuition and therefore it is convenient to recast the previous result using a more familiar language. To this end we analyze the concept of pure equilibrium states. If we consider a simple three-dimensional ferromagnetic Ising model with zero magnetic field at a temperature greater than the critical one, there is only one pure equilibrium state and the spins are disordered. Below the critical temperature there are two translationally invariant pure equilibrium states, one with positive magnetization the other with negative magnetization.

At temperatures below the roughening transition for the interface between regions of positive and negative magnetization, there are also pure equilibrium states, which are not translationally invariant; half of the space has positive magnetization and half of the space has negative magnetization. These new pure equilibrium states have much greater total free energy than the translationally invariant states. The difference in free energy is due to the presence of the interface; in the infinite volume limit the free energy density is the same, as it should be.

Generally speaking, a pure equilibrium state is any state which is at local equilibrium and whose connected correlation functions become zero at large distances. It is clear that the linear combination of two pure equilibrium states will be a mixed equilibrium state. According to conventional wisdom, a system, which is not at equilibrium at some initial time, evolves toward a pure equilibrium state and the time needed to go from one equilibrium state to another is exponentially large.

From the technical point of view, a pure equilibrium state is characterized by the clustering property; all the connected correlation functions become zero at large distances. Equivalently, intensive quantities do not fluctuate. In a nutshell, a pure equilibrium state corresponds to our intuitive idea of a normal equilibrium state. To avoid inconsistency with the thermodynamics, all equilibrium states must have the same free energy; states with higher free energy are metastable.

We have already seen that spin glasses may arrange the directions of their spins in many different ways and all these arrangements give similar values of the free energy, in other words, there exist many different ground states of spin glasses which are nearly degenerate in free energy. It is natural to assume that an Ising spin glass has many more equilibrium states than a ferromagnetic Ising model [3]. The configurations which have lower free energy correspond to pure equilibrium states while those configurations which have a higher free energy density in the infinite volume limit correspond to metastable states.

We now face the new problem of characterizing the ensemble of equilibrium states. The first thing to do is to assign an a priori probability w_α for the state α to appear in the ensemble. The natural choice is:

$$w_\alpha \propto \exp(-\beta F_\alpha), \tag{26}$$

where F_α is the free energy of the state α, although other possibilities are open.

The second task [6] is to study how all these states differ from one another; more precisely to define a distance between these states. A simple possibility for the distance between the state α and the state β is the following:

$$d^2_{\alpha\beta} = \frac{1}{N}\sum_{i=1}^{n}(m_i^\alpha - m_i^\beta)^2, \tag{27}$$

where N is the total number of spins, and m_i^α is the average magnetization of the spin i in the state α. In the same way we can define the overlap between two states α and β as

$$q_{\alpha\beta} = \frac{1}{N}\sum_{i=1}^{n} m_i^\alpha m_i^\beta . \tag{28}$$

If all the states have the same overlap with themselves (as should be the case in spin glasses), i.e. $q_{\alpha\alpha}=q_{\beta\beta}\equiv q_{EA}$ [4], the distance is very simply related to the overlap:

$$d^2_{\alpha\beta} = 2(q_{EA}-q_{\alpha\beta}). \tag{29}$$

It is possible to define the probability distribution of overlap P(q):

$$P(q) = \overline{P_J(q)} = \sum_{\alpha\beta} w_\alpha w_\beta \delta(q_{\alpha\beta}-q), \tag{30}$$

where the bar denotes the average over different samples with different value of the coupling J. In other words, P(q) is the probability of finding two states with overlap q, weighting each state with its probability of appearing in the ensemble. Only states which have a non zero probability ($w_\alpha \neq 0$) contribute to (30).

The physical interpretation of replica symmetry breaking is based on the result that the two functions P(q), defined in (16) and (30), coincide.

In the ferromagnetic Ising model at non-zero magnetic field, P(q) is a delta function because there is only one equilibrium state. Exactly at zero magnetic field the function P(q) contains two delta functions: one at $q=m^2$, the other at $q=-m^2$, where m is the spontaneous magnetization.

In spin glasses in the mean field approximation, the function P(q) at high temperatures is a delta function, at low temperatures at non-zero magnetic field has two delta functions with a smooth region in between. In this case, where the function P(q) is not a simple delta function, we say that the replica symmetry is broken. The transition is very smooth from the thermodynamical point of view: only the second derivative of the specific heat is discontinuous.

According to the established theory for spin glasses, all possible probability distributions with more than one overlap can be computed from the function P(q). The free energy can be computed as a functional of P(q) and the function P(q) can be found by looking for the maximum of free energy. The first interesting result is that $P_J(q)$ fluctuates from sample to sample [6]:

$$\overline{P_J(q_1)P_J(q_2)} = \tfrac{1}{2}P(q_1)P(q_2)+\tfrac{1}{2}\delta(q_1-q_2)P(q_1). \tag{31}$$

This result indicates that only a few states give an important contribution to the function $P_J(q)$ and their relative weight and overlap fluctuate from sample to sample.

The most interesting and unexpected result concerns the probability distribution of the overlap of three or more states. One finds that the distance satisfies the ultrametricity inequality [6]

$$d_{\alpha\beta} < \max(d_{\alpha\gamma}, d_{\beta\gamma}),\ \forall\gamma. \tag{32}$$

As is well known to mathematicians this inequality implies that the space of states can be divided into clusters of states, each cluster of states may be further divided into subclusters and so on. A cluster can be characterized as the set of states whose distance from a given, but arbitrary, state is smaller than a specified value.

A more intuitive definition of an ultrametric distance is that two spheres are identical if they have a point in common. As a consequence, if we consider a random walk in the space of the states of the system, and the length of each step is less than or equal to δ, even after M steps the distance between the initial point and the final point will be less than or equal to δ. The whole cluster can be reached in one step and there is nothing else to explore at distances less than δ.

This result is extremely important because it may form the basis for the existence of many different scales of relaxation times. Indeed let us consider a system which evolves in time and relaxes toward equilibrium with many different scales $\tau_1 \ll \tau_2 \ll \tau_3 \ll \ldots \ll \tau_n$. If the system is in a given configuration at time 0, we call R_k the region of configuration space that the system may explore in time τ_k. All the points in configuration space which belong to R_k but not to R_{k-1} are said to be at distance k.

It is evident that this definition of distance satisfies the ultrametricity inequality (32). Let us suppose the contrary, i.e. that there are two configurations (α and β) at distance k and there is another configuration (γ) at distance k-1 from both α and β: now the times needed to go from α to γ and from γ to β are both equal to τ_{k-1}; the estimated time for going from α to β via γ is then about $2\tau_{k-1}$, in contradiction with the inequality $\tau_{k-1} \ll \tau_k$.

This result is not surprising; many different times of relaxation are naturally present if the free energy landscape in configuration space contains many valleys which are separated by high mountains. The time to go from one valley to another is the exponential of the height of the lowest saddle between the two valleys.

The division of configuration space into valleys is clearly ultrametric. More precisely, if the distance between two configurations is defined as the height of the highest saddle one must cross in making the most convenient trip between the two configurations, this definition of distance satisfies the ultrametric inequality (32). The real surprise is that the space of equilibrium states of a spin glass is ultrametric using the natural definition of the distance.

The dynamics of a spin glass at finite temperature should be the following: at relative short time the system evolves exploring the configuration space corresponding to one pure equilibrium space; at much larger times (e.g. for times proportional to $\exp\{(Nd)^{1/4}\}$) the system may arrive at other equilibrium states at distance d from the original one. A careful study of the dynamics is very important [7]. Very interesting results have been obtained. In particular, it has been stressed that the behaviour of the system may be quite different if the system starts from a true equilibrium state or from a state which is slightly off equilibrium [8].

Another result which is very interesting is that the specific form of the equilibrium states is very sensitive to external parameters: a small change of the external magnetic field, of the order of $1/N^{1/2}$, is sufficient to completely upset the microscopic details of the equilibrium states: the equilibrium states at two different magnetic fields have very small overlap [3]. The magnetization as a function of the magnetic field makes many small jumps of the order of $1/N^{1/2}$. We could say that when we change the magnetic field, the system undergoes a sequence of first order microtransitions.

All these results certainly hold when fluctuations are neglected. The approximation is good when the dimensions of the spaces are very large, and the predictions of this approach compare well with the results of the numerical simulation in infinite dimensions, i.e. for the Sherrington and Kirkpatrick model [9].

If we decrease the dimensions, the situation is less clear [10,11,12]. The most likely scenario, among many infinitely different possibilities, is that the picture does not change for dimensions greater than D_c. At the lower critical dimension D_c, the critical temperature becomes zero and there is no transition to a spin glass phase. However if the coherence length ξ is large and if the relaxation time is proportional to $\exp(\xi^{1/4})$, it is possible that the relaxation time becomes very large (seconds, days, years...) and for reasonably long observation times the system seems to be in the spin glass phase.

In three dimensions the situation is confusing: many data on the coherence length ξ can be fitted as $(T-T_c)^{-\nu}$ with a reasonable value of T_c, but the same data can also be fitted with $\exp(A/T^3)$. Since corrections to simple scaling behaviour are also expected we cannot decide between the two different options by looking at the value of the χ^2 of the fit. It is difficult to find a critical test.

A very neat way of solving the problem consists in computing the corrections induced by fluctuations in the framework of the present theory and then comparing them to the experimental and numerical data. If the agreement is good, the value of the lower critical dimension should be the theoretical one. Unfortunately such computation is very difficult and progress seems slow in being achieved [13].

References

1) For a theoretical review on the mean field approximation in spin glasses see M. Mezard, G. Parisi and M. Virasoro, "Spin Glass Theory and Beyond" (World Scientific, Singapore, 1987).
2) For a nice short experimental review of spin glasses, see J.A. Mydosh, in "Disordered Systems and Localization", ed. by C. Castellani, C. Di Castro and L. Peliti (Springer-Verlag, Berlin, 1981).
3) G. Parisi, Phys. Rev. Lett. 43, 1754 (1979), J. Phys. A13, 1101; 1887; L115 (1980), Phys. Rev. Lett. 50, 1946 (1983).
4) S.F. Edwards and P.W. Anderson, J. Phys. F5, 965 (1975); F6, 1927 (1976).
5) J.R.L. de Almeida and D.J. Thouless, J. Phys. A11, 983 (1978).
6) M. Mezard, G. Parisi, N. Sourlas, G. Toulouse and M. Virasoro, Phys. Rev. Lett. 52, 1156 (1984); J. Physique 45, 843 (1984).
7) H. Sompolinsky and A. Zippelius, Phys. Rev. Lett, 50, 1297 (1983) and references therein.
8) A. Houghton, S. Jain, A.P. Young, Phys. Rev. B25, 2630 (1983),
9) D. Sherrington and S. Kirkpatrick, Phys. Rev. Lett. 35, 1792 (1975); Phys. Rev. B17, 4384 (1978).
10) A.P. Young, Phys. Rev. Lett. 51, 1206 (1983).
11) N. Sourlas, LPTENS preprint 85/9 (1985).
12) R.N. Bhatt and A.P. Young, Phys. Rev. Lett. 54, 924 (1985); A.T. Ogielsky and I. Morgenstern, Phys. Rev. Lett. 54, 928 (1985).
13) C. De Dominicis and I. Kondor, J. Physique Lett. 46 L-1037 (1985) and references therein.

Formation of Order and Patterns

K. Kawasaki
Department of Physics, Faculty of Science,
Kyushu University, Fukuoka 812, Japan

Nonlinear dissipative field theory of order formation processes in phase transitions is described. Considerable progress is made by the reduction of field theory to pattern dynamics. The approach is illustrated by ordering kinetics of alloys, layered magnets, and polymer blends.

1. Introduction

Phenomena of formation of order are very widespread in nature, ranging from what happened in the early universe [1] and the gravitational collapse [2] to a variety of processes that occur in laboratory experiments. Thus, a thorough understanding of the fundamentals of order-formation processes is of great importance in various areas of natural sciences. Naturally, the study of this problem has quite a long history and has many aspects [3]. An impressive accumulation of knowledge now exists, for instance, in the area of crystal growth which is a typical order-formation process. However our understanding of order formation is still far from complete. One source of the difficulties can be traced to the fact that here, in a sense, we are dealing with a kind of nonlinear field theory in which nonlinearity enters in an essential manner. Thus, in a way, we are facing similar problems to those faced in QCD (quantum chromodynamics). Indeed, we shall see that our starting equation is formally similar to the one that appears in the stochastic quantization of QCD [4].

In this article we shall discuss order formation in condensed matter physics as an example of nonlinear dissipative field theories. The relevance of such studies in theoretical physics, besides the insights gained into what is happening in individual order-formation processes in laboratory experiments, is that the studies provide us with examples of successful attempts (and also unsuccessful attempts, which will never be mentioned) to tackle nonlinear field-theoretical model systems. One obvious advantage of working in condensed matter physics in this connection is existence of a wealth of real systems for which high quality experimental information is becoming available. This fact, in conjunction with availability of computer simulation studies, is certainly of enormous help in constructing theories. S. Tomonaga [5] once wrote: "In a way a physicist is among the least smart. In solving an exam problem he cheats and finds an answer by secretly doing an experiment. Otherwise he is often never able to understand".

Whether or not one calls doing an experiment cheating, it is true that those least smart physicists were often able to create, with the help of experimental information, a respectable body of knowledge which often surpasses in its depth and influence what smarter people could achieve. A recent prototype example is provided by our understanding of critical phenomena where the idea of universality was born and the method of renormalization group was developed. Here we should be reminded of the fact that these developments were motivated to a large extent by the enormous amount of experimental information on critical behavior of many systems accumulated since the end of the last century [6].

2. Field-Theoretical Models and Early Theories

Broadly speaking, there are two classes of order-formation process associated with systems quenched from a high temperature disordered state into low temperature thermodynamically metastable and unstable states within the coexistence curve in the phase diagram schematically shown in Fig. 1. During approach to thermal equilibrium following a quench into a metastable state there is an energy barrier to be overcome which is much greater than $k_B T$. This process which is called a nucleation process is generally very slow and starts with the formation of spherical droplets called nuclei by thermal activation. For a quench into an unstable state, on the other hand, no discernible energy barrier much greater than $k_B T$ exists, and fluctuations start to develop very rapidly after the quench.

The field-theoretical treatments of nucleation processes are already well developed [7], and we shall not discuss these in this article. Thus here we will be mainly concerned with what happens after a system is quenched into an unstable state. Our first task is to construct a field-theoretical model. For this purpose here we adopt a phenomenological approach and start from the thermodynamics of irreversible processes. The time variation of the order parameter S(t) is then proportional to the thermodynamic driving force for change which is the derivative H'(S) of the thermodynamic potential H(S). Thus we have

$$\frac{d}{dt} S(t) = -LH'(S(t)) \tag{2.1}$$

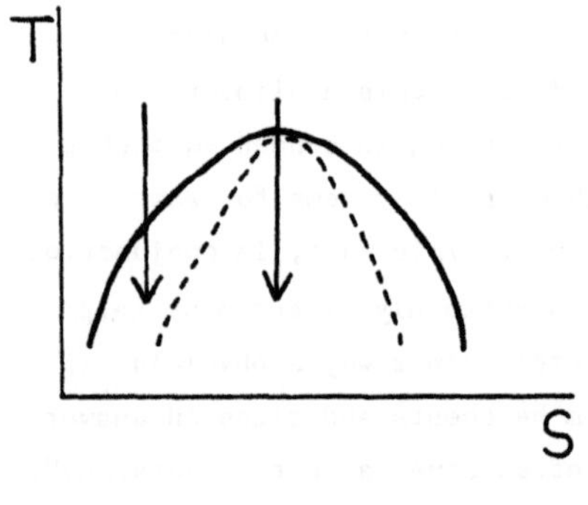

Fig.1 The phase diagram in the S(order parameter)-T(temperature) plane

where L is a positive constant called the Onsager kinetic coefficient. Typically H(S) takes the following Landau form:

$$H(S) = -\frac{1}{2}\tau S^2 + \frac{1}{4!}gS^4 \tag{2.2}$$

where τ and g are positive constants. The negative sign in front of the first term expresses thermodynamic instability which gives rise to exponential growth of a very small initial fluctuation S(0);

$$S(t) = S(0)\exp(tL\tau). \tag{2.3}$$

The model (2.1) is oversimplified in that it has no spatial structure and is also incapable of accommodating the cases with the conservation law, that is, the invariance of $\int S(\underline{r},t)d\underline{r}$. These can be taken care of by first replacing (2.2) by the Ginzburg-Landau free energy functional

$$H\{S\} = \int d\underline{r}\{\frac{1}{2}\tau\xi_0^2[\underline{\nabla}S(\underline{r})]^2 - \frac{1}{2}\tau S(\underline{r})^2 + \frac{1}{4!}gS(\underline{r})^4\}, \tag{2.4}$$

and by replacing H'(S) by the functional derivative $\delta H\{S\}/\delta S(\underline{r})$ and L by $L(-\nabla^2)^\alpha$ where $\alpha=0$ or 1 depending on absence or presence of the conservation law saying that the integral of $S(\underline{r})$ over space remains constant. Here τ and g are positive constants and ξ_0 is some microscopic length. The final outcome is the following time-dependent Ginzburg-Landau equation:

$$\frac{\partial}{\partial t}S(\underline{r},t) = -L(-\nabla^2)^\alpha\delta H\{S(t)\}/\delta S(\underline{r},t). \tag{2.5}$$

For completeness we must further add a random force $f(\underline{r},t)$ on the r.h.s. of (2.5) which obeys the fluctuation-dissipation relation to include effects of thermal noise. This is required to guarantee approach to the equilibrium state described by the equilibrium distribution functions of the order parameter proportional to $\exp(-H\{S\}/k_BT)$. This, however, will be omitted since here we will be concerned with situations where the thermal noise does not play a role (which is not the case for nucleation).

Now, the dashed line in Fig. 1 dividing the region inside the coexistence curve into thermodynamically metastable and unstable regions, which is called the spinodal line, is given by

$$H''(S) = 0 \qquad \text{(spinodal line)} \tag{2.6}$$

and depends on the values of the parameters τ and g. It is important to realize, however, that these parameters cannot be assigned unique values reflecting the fact that the G-L functional (2.4) itself depends on the short-distant cut-off which is inevitably associated with any continuum model [8]. Thus the location of the spinodal line as well as the distinction between metastable and unstable states depend on the short length scale with which we want to describe the system (renormalization effects). Indeed such cut-off dependence of the location of the spinodal line was investigated by Monte Carlo technique [9] and by renormalization group [10]. We will not pursue this problem further in the following.

In this article we only consider the so-called critical quench where the average order parameter $\langle S(\underline{r},t)\rangle$ is maintained to be zero. Then, immediately after a quench, the local order parameter fluctuation $S(\underline{r},t)$ is very small and we may linearize (2.5). We then find that the spatial Fourier component $S_{\underline{k}}(t)$ of $S(\underline{r},t)$ is shown to behave as

$$S_{\underline{k}}(t) = S_{\underline{k}}(0)\ \exp(t\Gamma_{\underline{k}}) \tag{2.7}$$

with the growth rate given by

$$\Gamma_k = L\ \tau\ k^{2\alpha}(1-\xi_0^2\ k^2) \tag{2.8}$$

where $S_{\underline{k}}(0)$ is a fluctuation which existed at the time of quench. Thus the growth due to thermodynamic instability occurs only for wave numbers smaller than some threshold: $k < \kappa \equiv \xi_0^{-1}$. In particular, for conserved cases $\alpha=1$, the growth rate reaches a peak value at $k=k_m \equiv 2^{-1/2}\ \kappa$. This is an outline of the linear theory of growth due to Cahn and Hilliard [11] and early experimental data were analyzed in terms of this theory. While this theory correctly relates the growth of fluctuations to thermodynamic instability, the theory is grossly inadequate as a quantitative theory. Indeed, it is only very recently that examples were found where the linear theory was experimentally confirmed for the case of polymers with very long time constants [12]. For most cases growth occurs so rapidly that phase separation is completed locally in very short times of typically several times $\xi_0^{2\alpha}/L\tau$. (The argument for this is briefly outlined in Appendix.) That is, the absolute value $|S(\underline{r},t)|$ is near to its saturation value $M \equiv (6\tau/g)^{1/2}$ obtained by minimizing (2.2).

Important evidence against the linear theory comes from the discovery of a scaling law [13] which is expressed as

$$I_k(t)\ \equiv \langle|S_{\underline{k}}(t)|^2\rangle/\int d\underline{k}'\langle|S_{\underline{k}'}(t)|^2 = \ell(t)^d F[k\ell(t)] \tag{2.9}$$

where $\ell(t)$ gives a characteristic length scale for fluctuations, F is a scaling function which no longer depends explicitly on time, and d the dimensionality of space. This law appears to be widely confirmed except for the initial transient regime. We thus realize that the nonlinear term with g in (2.5) plays an essential role. Despite ingeneous attempts, no quantitatively satisfactory theory exists which consistently treats both the early linear and late nonlinear regimes. Of course, nonlinear problems of this sort are common in various branches of physics. The outstanding example where such nonlinearity was successfully dealt with non-perturbatively is the renormalization group theory of critical phenomena. Naturally we find attempts to apply renormalization group ideas to the order-formation kinetics [14]. Although there remains hope that this line of approach will eventually solve the problem, we have not yet seen any decisive work. A root of the difficulty of our problem as compared to that of critical phenomena lies in the greatly enhanced level of fluctuations. Critical phenomena are associated with thermodynamical marginal stability where thermal fluctuations get enhanced due to weakening of restoring forces to suppress fluctuations. On the ofter hand, for quenched systems we start with a state at the hill top of the thermodynamic potential. Thus amplitude of fluctuation in $S(\underline{r},t)$ can grow to the maximum M which is well beyond the thermal level and is much greater than that of critical phenomena at least for higher dimensionality (greater than 3) where the expansion from 4 dimensions is meaningful.

3. Pattern Dynamics

The difficulty of incorporating nonlinearity into the field equation (2.5) compels us to put aside the problem temporarily and to search for some alternative approaches. A clue is found in what we have already said in the preceding section, i.e., that after a short time, $S(\underline{r},t)$ is almost equal to M or -M. That is, we have a domain structure after a short transient time. The driving force $-\delta H/\delta S$ then vanishes almost everywhere except on domain boundaries (interfaces) whose configuration is the only degree of freedom left. This then suggests the possibility of converting the original field equation (2.5) into an equation of motion of domain walls. This is a typical example of information reduction in statistical physics. Other more familiar examples are the derivation of the Boltzmann equation of gas dynamics from the Newton or Liouville equation for a system of 10^{26} gas molecules, and derivation of the hydrodynamic equations from the Boltzmann equation. A successful reduction usually presupposes the existence of a smallness parameter. For the Boltzmann equation this parameter is the ratio of molecular diameter to mean free path. For hydrodynamics, it is the ratio of the mean free path to the macroscopic length. Thus here we have a two-step reduction in going from microscopic to macroscopic behavior. In our case such a parameter is provided by the ratio of the domain wall width to the characteristic length scale of domain walls e.g. the average radius of curvature of domain walls.

Here we must point out that the theory of order formation in terms of domain walls is not new [15]. In fact, the theory of nucleation and growth, and also that of grain growth and crystal growth have always dealt with the motion of phase boundaries. Here we attempt to build a unified theory that will provide a foundation and will also extend the range of applicability of the approach to such a problem as the behavior of strongly correlated fluctuations in quenched systems including complex systems like polymers. A new element in the old interface equations of motion is addition of the effects of thermal noise $f(\underline{r},t)$ neglected in (2.5), of which there are two kinds. One possibility is to add random force terms obeying fluctuation-dissipation relations to the deterministic interface equations like (3.1) and (3.2) below, which can be done quite generally [16,17]. Another is to include processes in which interfaces are randomly created by thermal fluctuations as in nucleation processes, for which we do not yet have a general theory. In this article, however, we do not discuss this problem.

Reduction of the field equation like (2.5) can be done by various techniques such as projection methods [18] or multi-scale perturbation methods [19], which are more or less standard and are briefly explained in the present context in Appendix. Here we present reduced interface equations only for the two typical cases without the conservation law ($\alpha=0$).

The first example is for (2.5) with $\alpha=0$ and spatial dimensionality greater than one. The resulting interface equation takes the form [15,16]

$$v(a) = L\,\tau\xi_0^2\,K(\underline{a}) \tag{3.1}$$

where $v(\underline{a})$ is velocity of the interface in the normal direction and $K(\underline{a})$ the mean curvature in which $\underline{a}$ parametrizes the interface. The appearance of mean curvature here is readily understood since a change of interfacial area due to infinitesimal interfacial displacement along the normal direction is proportional to $K(a)$ which provides a driving force for interfacial motion.

Another example is for the same model except that there is a strong anisotropy in $H\{S\}$ such that the interfaces always orient themselves perpendicularly to the x-axis and the problem becomes one dimensional. In this case the driving force proportional to $K(a)$ is absent and much smaller driving forces neglected in (3.1) will appear which come from a weak exponential attractive potential between neighboring flat interfaces. The equation of motion is then expressed in terms of x_i, the x coordinate of the intersection of the i-th domain wall and the x-axis, where flat interfaces are consecutively numbered from left to right along the x-axis. Thus we have [20]

$$\frac{d}{dt}x_i = e^{-(x_{i+1}-x_i)/\xi} - e^{-(x_i-x_{i-1})/\xi} \tag{3.2}$$

where the coefficient in front is absorbed into t, and ξ is the domain wall width.

If we are interested in time evolution of simple interface configurations, the equations like (3.1) and (3.2) can be directly solved to obtain the desired results. However, in ordering kinetics we are often interested in complicated interface patterns where only some average properties like the fluctuation spectrum (2.9) are of physical interest. We may be also interested in macroscopic behavior to be described by macroscopic equations like hydrodynamic equations of systems finely divided into small domains as exemplified by microemulsions or composite materials with microstructure. Then, the second-step reduction of mesoscopic kinetic equations of patterns like interface equations (3.1) and (3.2) is necessary, which is analogous to the derivation of hydrodynamics from the Boltzmann equation in gas dynamics. However, the task here is more difficult due to the absence of a general expansion technique using a smallness parameter which is the ratio of the average domain size to the macroscopic length, corresponding to the Enskog-Champan method in gas dynamics. Therefore, such a method must be invented for each problem. Here we merely outline the two cases where this was done.

The first is the case described by (3.1) and is found in [21]. Since this equation is still highly nonlinear and is not generally tractable, a crucial step is to convert it into the equation for an auxilliary field $u(\underline{r},t)$ which vanishes only on the interfaces as (the arguments $\underline{r}$, t of u are suppressed)

$$\frac{\partial}{\partial t} u = L\tau\xi_0^2 (\nabla^2 - \sum_{ij} n_i n_j \frac{\partial^2}{\partial r_i \partial r_j})u \tag{3.3}$$

with $n_i = |\underline{\nabla} u|^{-1} \partial u/\partial r_i$. This equation is still highly nonlinear. However, if we are interested in states with highly complex random interface configurations which are statistically isotropic and homogeneous, (3.3) can be linearized by approximating $\underline{\underline{nn}}$ by $d^{-1}\underline{\underline{1}}$ where d is the spatial dimensionality and $\underline{\underline{1}}$ is the unit tensor. This reduces (3.3) to the diffusion equation for u with the diffusion constant $D \equiv (1-d^{-1}) L\tau\xi_0^2$. If we further assume that $u(\underline{r},t=0)$ as a random field obeys the Gaussian distribution with a microscopic correlation length, the problem is completely solved to yield the following closed form expression for correlation of order parameter fluctuations;

$$\langle S(\underline{r},t)\, S(\underline{r}',t)\rangle / \langle S(\underline{r},t)^2\rangle = (2/\pi)\sin^{-1}\exp[-(\underline{r}-\underline{r}')^2/8Dt] \tag{3.4}$$

This expression is seen to conform to the scaling (2.9) with $\ell(t)=(Dt)^{1/2}$. This result is favorably compared with those of experiments and computer simulations. The approximation is expected to be better for higher spatial dimensionality. It is interesting to observe that the growth law is written roughly as $\ell^2/t \sim \xi_0^2/t_0$ with $t_0=1/L\tau$ the time of local phase separation. Hence ξ_0 gives the size of domains during the initial transient period.

Another example is the ordering kinetics in quasi-one-dimensional systems like layered magnets [22]. Here we can start from (3.2) as a first approximation. Although this equation cannot be analytically solved for cases with large numbers of domain walls (or kinks and antikinks), it is easily solved by computer provided that any kink-antikink pair within the distance ξ is made to annihilate [23]. In this way we can find the average domain size and their distribution. The results are summarized as follows: (i) scaling behavior was confirmed after a short initial transient with the characteristic length $\ell(t)$ growing logarithmically in time, and (ii) the scaled domain size distribution exhibits a sharp cut-off structure with no domains with sizes below the cut-off. Reflecting these features the scaled scattering structure function (SSF) $I_k(t)$ obeys the scaling law (2.9) in agreement with experimental results of Ikeda [22], but the scaling function F(z) decreases rapidly far below the experimental results as z increases [24]. The latter disagreement is traced to the fact that flat domain walls are in fact unstable against undulatory deformations in the presence of attractive domain wall interactions. Therefore the starting model (3.2) must be generalized. When this is done, the result is given again as a closed form expression for the scaled SSF with a logarithmically increasing characteristic length [24]. The agreement of our theory with Ikeda's experimental result turned out to be quite remarkable with no adjustable parameter.

The two cases described above are the only known cases with closed form expressions for the scaled order parameter correlation functions of interconnected domain wall states of ordering kinetics. Such closed form expressions could not have been easily obtained by means of other than the interface dynamics described in this section.

4. Beyond TDGL and Polymer Blend

Up to now we have used as our starting point the phenomenological TDGL model equation (2.5) which was derived from general principles such as symmetry and conservation laws. Such a continuum model presupposes the existence of a length much greater than a microscopic length like lattice constant. Hence, rigorously speaking, the validity of such a continuum model is restricted to special cirmcumstances as found for example in critical phenomena or superconductivity. However, in ordering kinetics, such a model has often been used beyond its limit of validity as we have done. Therefore, the results obtained from this model must sometimes be treated with caution when they are compared with the results of laboratory experiments or computer simulations on discrete models. This seems to be the case especially when the order parameter is conserved. Here the value of the exponent of the growth law $\ell(t) \sim t^{\mu}$ is still highly controversial [13] although the continuum model predicts $\mu=1/3$.

On the other hand, even when the continuum treatment is valid, this does not always justify the use of a simple TDGL equation such as (2.5). For instance, the coefficient L can depend on the local order parameter and the equation can be nonlocal in space and can contain memory effects. A concrete example of this more complicated kind of field equation is the one which describes phase separation of a polymer blend, i.e. a mixture of two kinds of polymer chains, one consisting of N_A monomers of species A and another containing N_B monomers of species B. The characteristic length that permits continuum treatment is the radius of gyration R_G of a polymer chain $\propto N_A^{1/2}$ or $N_B^{1/2}$. Here the order parameter, aside from a constant, can be taken to be the number density ρ_A of monomers of species

A if we assume incompressibility i.e. $\rho_A+\rho_B=\rho_0$=constant, where ρ_B is the monomer density of the species B. Dynamics of polymer systems is one of the most complex many-body problems because of strong entanglements among polymer chains [25] and thus we must adopt a simplified model to make progress. A reasonably successful model is the reptation model put forward by de Gennes, Edwards and Doi [25] where a polymer chain can move randomly along a tube which represents constraints on a chain due to entanglements.

On the basis of this reptation dynamics, a continuum field model for the order parameter was recently constructed by the present author and K. Sekimoto [26]. The equation which neglects memory effects takes the following form with J=A or B:

$$\frac{\partial \rho_J(\underline{r})}{\partial t} = -\frac{3D_J N_J}{T^* b^2} \int [p_J(\underline{r})\delta(\underline{r}-\underline{r}') - P_J(\underline{r},\underline{r}')]\mu_J(\underline{r}')d\underline{r}'. \tag{4.1}$$

Here D_J is the self-diffusion constant of a J-type chain in a pure polymer system of species J, b is the length of a single chain segment (more precisely, a segment of the Doi-Edwards primitive chain), T^* is the absolute temperature times the Boltzmann constant, $p_J(\underline{r})$ is the number density of ends of polymer chains, and $P_J(\underline{r},\underline{r}')$ is the joint probability density that the ends of a chain are located at $\underline{r}$ and $\underline{r}'$ normalized as

$$\int P_J(\underline{r},\underline{r}')d\underline{r}' = p_J(\underline{r}). \tag{4.2}$$

Finally $\mu_J(\underline{r})$ is given by

$$\mu_J(\underline{r}) = \frac{\delta}{\delta\rho_J(\underline{r})} H\{\rho_A,\rho_B\} + \mu^*(\underline{r}) \tag{4.3}$$

where $H\{\rho_A,\rho_B\}$ is the coarse-grained free-energy functional and $\mu^*(\underline{r})$ is determined in such a way that (4.1) is compatible with the incompressibility condition. We see that (4.2) in fact guarantees the conservation law by integrating (4.1) over $\underline{r}$. $P_J(\underline{r},\underline{r}')$ and hence $p_J(\underline{r})$ can be determined as a functional of ρ_A and ρ_B by making use of the local equilibrium approximation which is consistent with the neglect of memory effects in (4.1). In this way we obtain (4.1) as a closed non-linear integral equation for $\rho_J(\underline{r},t)$. An important point to note is that the incompressibility condition inevitably introduces correlations between reptative motions of polymer chains.

The task of solving (4.1) is much more formidable than (2.5). Nonetheless it is possible to develop an interface dynamics starting from (4.1). In the strong segregation limit where phase-separated domains consist of almost pure phases, the interface width ξ is rather small and is independent of N_J. The interface equation then takes the form,

$$\frac{T^* b^2 \rho_0^2}{2D_J N_J^2} [\gamma^{-1} v(\underline{a}) + \int G(\underline{aa}') v(\underline{a}') d\underline{a}'] = \sigma\, K(\underline{a}) \tag{4.4}$$

where σ is the interfacial tension and

$$\gamma \equiv \int dz [\sum_J \frac{N_J}{p_J(z)}]^{-1} \tag{4.5}$$

$$G(\underline{aa}') \equiv \langle \underline{a}, z=0 \mid \sum_J N_J (p_J P_J^{-1} p_J - p_J)^{-1} | \underline{a}', z' = 0 \rangle . \tag{4.6}$$

Here z is the local coordinate perpendicular to the interfacial element at $\underline{a}$ and a matrix notation with indices $\underline{r}=(\underline{a},z)$ and $\underline{r}'=(\underline{a}',z')$ has been used in (4.6). The first term on the l.h.s. represents interface reaction kinetics where friction is localized on interfacial regions, whereas the second term arises from diffusion occurring well inside domains. Analysis of (4.4) is of course much more difficult than (3.1) and (3.2). However the following order of magnitude estimates are useful: $\gamma \sim p\xi/N$ and $G \sim N/pR_G^2 \ell^{d-1}$ where ℓ is the characteristic length of domain structure. Thus we see that for $\ell \lesssim \ell_c \equiv R_G^2/\xi$ the first term on the l.h.s. of (4.4) dominates, that is the conservation law is not effective. In this case (4.4) reduces to (3.1) and the characteristic length grows as $\ell(t) \propto t^{1/2}$. In fact there are experimental indications which might suggest the predicted behavior [27].

5. Concluding Remarks

In this article we have demonstrated how the pattern dynamic equations reduced from field theory models are effective in tackling difficult problems of order formation processes. This is but one example illustrating the importance of information reduction in dealing with complex problems having a hierarchical structure characterized by existence of widely separated scales. This need not be limited to problems in condensed matter physics. Derivation of QCD from superstring models may well be another example of this sort.

Appendix

Here we briefly analyze short and long time behaviors of (2.5) which is written out as

$$\frac{\partial}{\partial t} S(\underline{r},t) = -L\tau(-\nabla^2)^{\alpha}\{-\xi_0^2\nabla^2 S(\underline{r},t) - S(\underline{r},t)[1-\frac{S(\underline{r},t)^2}{M^2}]\}. \tag{A.1}$$

We first rescale this equation by introducing

$$S(\underline{r},t) = Ms(\underline{R},T) \tag{A.2}$$

with

$$\underline{R} \equiv \underline{r}/\xi_0 , \qquad T \equiv (L\tau/\xi_0^{2\alpha})t \tag{A.3}$$

to obtain

$$\frac{\partial}{\partial T} s(\underline{R},T) = -(-\nabla_R^2)^{\alpha}\{-\nabla_R^2 s(\underline{R},T) - s(\underline{R},T)[1-s(\underline{R}T)^2]\}. \tag{A.4}$$

Eq. (A.4) contains no parameter. Thus, shortly after the quench, growth of the order parameter will be governed by the length scale ξ_0 and the time scale $\xi_0^{2\alpha}/L\tau$. During this period the order parameter fluctuations with the amplitudes of order M will develop. After this period, the r.h.s. of (A.4) will be sharply reduced and the slower process of domain growth will start to take place with the much greater length scale than ξ_0 which typically behaves as $\xi_0 T^z$ with z some positive growth exponent. To see this more explicitly we take up the case without conservation $\alpha=0$. At late times $T \gg 1$ we will have well-defined domain wall states where time variation of the local order parameter is practically limited to the vicinity of domain walls. In any such vicinity spatial variation of $s(\underline{R},T)$ will take place primarily in the direction $\underline{n}$ perpendicular to the wall. This means that $s(\underline{R},T)$ is written as $s_0(u(\underline{R},T))$ where $s_0(u)$ is a definite function to be determined below and $u(\underline{R},T)$ is a function which vanishes for $\underline{R}$ just at the center of any interfacial region. The normal direction $\underline{n}$ is then given by

$$\underline{n} = \underline{\nabla}u/|\underline{\nabla}u| = \underline{\nabla}u \tag{A.5}$$

where suffixes R on u are omitted and u has been chosen such that $|\underline{\nabla}u|=1$. Using the fact that the (dimensionless) mean curvature of the domain wall $\bar{K}$ is given by $-\underline{\nabla}\cdot\underline{n}=-\nabla^2 u$, we find

$$\nabla^2 s_0(u) = -\bar{K}\, s_0'(u) + s_0''(u) \tag{A.6}$$

where primes on s_0 denote differentiations. Then (A.4) with $\alpha=0$ becomes

$$\frac{\partial u}{\partial T} s_0'(u) = -\bar{K}\, s_0'(u) + s_0''(u) + s_0(u)[1-s_0(u)^2]. \tag{A.7}$$

$s_0(u)$ is now chosen to satisfy the equation

$$s_0''(u) + s_0(u)[1-s_0(u)^2] = 0. \tag{A.8}$$

Thus $s_0(u)$ is just the equilibrium order parameter profile near a flat domain wall with u the dimensionless coordinate perpendicular to the wall. (A.7) then gives

$$-\frac{\partial u}{\partial T} = \bar{K} \tag{A.9}$$

which is also regarded as the solvability condition of the partial differential equation (A.4) with $\alpha=0$. Since $-\partial u/\partial T$ is the dimensionless velocity $\bar{v}$ of a moving domain wall along its normal direction, (A.9) is the dimensionless form of (3.1). Since the dimensionless mean curvature is generally much smaller than unity, the dimensionless time scale associated with (A.9) is much greater than unity. The multiscale perturbation methods [19] formalize the idea outlined here where the small perturbation parameter is $K\xi_0$. The projection method, on the other hand, makes use of the fact that the order parameter variation or the l.h.s. of (A.7) associated with the domain wall motion is equal to $-\bar{v}s_0'(u)$. Thus $\bar{v}$ can be extracted by multiplying both sides of (A.7) by $s_0'(u)$ and integrating them over u.

References

[1] T. Yoshimura, in this proceedings
[2] T. Nakamura, in this proceedings
[3] J.D. Gunton, M. San Miguel and P.S. Sahni, in "Phase Transitions and Critical Phenomena", vol.8, eds. C. Domb and J.L. Lebowtiz, Academic Press, New York (1983)
[4] G. Parisi and Y. Wu, Scientia Sinica 24, 483 (1981).
[5] S. Tomonaga, "What is physics all about" Iwanami Syoten, Tokyo (1979)
[6] J.M.H. Levelt Sengers, Physica 82A, 319 (1976)
[7] J.S. Langer, Ann. Phys. (N.Y.) 41, 108 (1967), ibid. 54, 258 (1969)
[8] J.S. Langer, Physica 73, 61 (1974)
[9] K. Kaski, K. Binder and J.D. Gunton, Phys. Rev. B29, 3996 (1984)
[10] K. Kawasaki, T. Imaeda and J.D. Gunton, in "Perspectives in Statistical Physics" ed. H.J. Raveche, North-Holland, Amsterdam (1981)
[11] J.W. Cahn and J.E. Hilliard, J. Chem. Phys. 28, 258 (1958)
[12] T. Hashimoto, J. Kumaki and H. Kawai, Macromolecules 16, 641 (1983)
H.L. Snyder, P. Meakin and S. Reich, Macromolecules 16, 757 (1983)
[13] Reviewed, for instance, by H. Furukawa, Advances in Phys. 34, 703 (1985)
[14] G.F. Mazenko, O.T. Valls and F.C. Zhang, Phys. Rev. B31,4453 (1985), and earlier references quoted.
J. Vinals, M. Grant, M. San Miguel, J.D. Gunton and E.T. Gawlinski, Phys. Rev. Lett. 54, 1264 (1985)
[15] S.M. Allen and J.W. Cahn, Acta Metall. 27, 1085 (1979)
[16] K. Kawasaki and T. Ohta, Prog. Theor. Phys. 62, 147 (1982)
R. Bausch, V. Dohm, H.K. Janssen and R.K.P. Zia, Phys. Rev. Lett. 47, 1837 (1981)
[17] K. Kawasaki and T. Ohta, Physica 118A, 175 (1983)
[18] K. Kawasaki Ann. Phys. (N.Y.) 154, 319 (1984)
[19] A.H. Nayfeh, "Perturbation Methods", Wiley, New York (1973)
[20] K. Kawasaki and T. Ohta, Physica 116A, 573 (1982)
[21] T. Ohta, D. Jasnow and K. Kawasaki, Phys. Rev. Lett 49, 1223 (1982)
[22] H. Ikeda, J. Phys. C. Solid State Phys. 16, 3563 (1983), ibid 19 L535 (1986)
[23] T. Nagai and K. Kawasaki, Physica 120A, 587 (1983)
[24] K. Kawasaki and T. Nagai, J. Phys. C. Solid State Phys. 19, L551 (1986)
[25] M. Doi and S.F. Edwards, "The theory of polymer dynamics" Oxford University Press, Oxford (1986)
[26] K. Kasawaki and K. Sekimoto, papers to be published in Physica A and in Europhys. Lett.
[27] T. Izumitani, M. Takenaka and T. Hashimoto, Polym. Prepr. Jpn. 35, 2974, 2978 (1986)

Index of Contributors

MIX
Papier aus verantwortungsvollen Quellen
Paper from responsible sources
FSC® C105338

If you have any concerns about our products,
you can contact us on
ProductSafety@springernature.com

In case Publisher is established outside the EU,
the EU authorized representative is:
Springer Nature Customer Service Center GmbH
Europaplatz 3, 69115 Heidelberg, Germany

Printed by Libri Plureos GmbH
in Hamburg, Germany